SYSTEM DYNAMICS

Theory and Case Studies

SYSTEM DYNAMICS

Theory and Case Studies

Second Edition

B. J. Ranganath

M. Sc. (Engg.), Ph.D.

Professor & Head

Dept. of Mechanical Engineering, VVIET, Mysore, Karnataka

Lewlyn L. R. Rodrigues

B.E., MBA, M. Tech., Ph.D.

Professor & Head

Humanities & Social Sciences

Manipal Institute of Technology

Manipal University, Manipal, Karnataka

TECHSAR™

NEW DELHI

Published and Printed by
TechSar Pvt. Ltd.
4435-36/7, Ansari Road, Daryaganj
New Delhi-110 002 (India)
E-mail: tech_info@techsarworld.com
Website: www.techsarworld.com

ISBN 978-93-81141-81-6

Reprint 2025

Preface to the Second Edition

System Dynamics is an emerging field in the study and analysis of a working system. This includes a wide variety of areas ranging from industrial manufacturing and management, service sectors like insurance, corporate planning, etc. to biological and medical research. This book is the outcome of the research of the authors in the basic areas of working systems in different areas of manufacturing, industrial and business management. It includes the experience they had in solving many real-life problems across many organizations.

Even though the focus here is on Engineering/Technology Management, the principles of developing dynamic hypothesis, building causal relations, constructing stock and flow diagrams and performing simulation are useful for any real-life situations, which involve dynamic analysis in one form or the other. A systematic approach for developing a System Dynamics model is detailed, which is helpful in defining and simulating the problem for arriving at a feasible solution.

Independent chapters on HRD, project management, new product development and robotic arm simulation deal with the systematic approach for moving towards optimal solution. Special case studies with reference to National Engineering Competence Pool Development and Deployment (NECPDD) and Insurance sector dynamics are highlighted in the book. Each chapter is presented in a systematic manner with Objectives, Problem Statement, Methodology, Summary and Exercise for better understanding of the topics. This makes the book a comprehensive resource material to System Dynamics for students and researchers at university level, and practising managers in the industries to develop their Project Models of interest and get feasible solutions.

In this second edition, new chapters have been included in innovation management, supply chain management and robotic arm simulation and thus giving a wider coverage of applications of System Dynamics which keeps growing forever. We are confident that the readers of this book will be able to get a thorough exposure to the basic concepts of System Dynamics as well apply them in their specific areas of interest.

B. J. Ranganath
Lewlyn L. R. Rodrigues

Acknowledgements

At the outset, we thank the inventor of the field System Dynamics J.W. Forrester. If not for his unparalleled insight into this area of science and technology, we would not have been able to write a book on this subject that has gained popularity particularly during the last decade.

This book, except for the first two chapters, is the narration of real life case studies which have been undertaken as the project in industries by our students. So, we wish to place on record the contribution of our research scholars and students some of whom are now our colleagues. We wish to place on record the contribution of Dr. Gopalakrishna, Mr. Dharmaraj, Mr. Vasanth, Mr. Rakesh Patel and Mr. Bharath Kantharaj, the students who worked meticulously on the projects assigned to them and contributed chapters to this book. We also thank the industries which have permitted us to collect the data for building the models and offered projects to our students through which we could build the models discussed in this book.

Last but not the least, we sincerely thank all our readers of the first edition, who have not only appreciated but also constructively criticized our work, which has enabled us to bring out this second revised edition.

B. J. Ranganath
Lewlyn L.R. Rodrigues

Contents

Authors

Dr. **B. J. Ranganath** has over four decades of teaching and research experience. Having obtained his Ph.D. from IIT-Madras he had an extensive service at National Institute of Engineering, Mysore and now at Vidya Vikas Institute of Technology, Mysore. His areas of research include Metal Cutting, Tool Design, Material Sciences, Manufacturing and several allied technological fields. He has guided several B.E. and M.Tech. projects and Ph.D. scholars and has over 125 publications in national and international conferences and journals. He is also an acclaimed writer on Spirituality and hundreds of students and general readers are benifitted immensely from his in-depth knowledge in Science, Technology and Spiritualism. Dr. B. J. Ranganath may be contacted at bjranganath@yahoo.co.in.

Dr. Lewlyn L. R. Rodrigues is a passionate researcher in System Dynamics. He started his teaching career in 1984 and has taught Engineering and Management courses in India and abroad. Currently, he is the Professor and Head of Humanities & Social Sciences at Manipal Institute of Technology, Manipal. He is B.E. in Mechanical Engineering, MBA in HR & Quality, M.Tech. in Production Engineering and Ph.D. in System Dynamics. He has guided several B.E., MBA, M.Tech. projects and successfully co-guided two Ph.Ds. He is currently guiding eight Ph.Ds. and has about 100 papers in conferences and reputed journals. His areas of research interest include System Dynamics, Total Quality Management, Knowledge/Technology Management, Manufacturing, and Human Resource Management. Dr. Rodrigues may be contacted at rodrigusr@gmail.com.

1

System Dynamics: Introduction

Objectives

After going through this chapter you will be able to:

- Define System Dynamics.
- Understand the concept of Complex System.
- Know the meaning of System Behaviour.
- Appreciate the wide range of applications of System Dynamics.
- Have a clear understanding of the basic definitions in System Dynamics.

1.1 BACKGROUND

System Dynamics, which basically deals with the simulation of interactions between objects in dynamic systems, was introduced by J. W. Forrester way back in late 50s. The systematic methodology suggested by him, was used thereafter, in a wide variety of applications including: Corporate planning and policy design, Biological and medical modelling, Energy and environment studies, Natural and social sciences, Decision-making, Complex non-linear dynamics, etc. Today, System Dynamics has developed itself into a powerful tool that finds its applications in almost all walks of life, where the behaviour of a system is to be studied.

1.2 WHAT IS SYSTEM DYNAMICS?

When we refer to a 'dynamic system' it means that its 'state' changes with 'time'. For a production engineer it may be the dynamics involved in a metal cutting process, for a mathematician it may be the dynamics of random number generation, for a biologist it may be a way how a disease propagates, for a businessman it may be the way how price of a product changes, for a manager it may be the fluctuations in the market, for an environmentalist it may be the way how weather changes, for an ecologist it may be the way in which the plants evolved from their aquatic origins, and so on. Hence, System Dynamics is a multidisciplinary subject, which basically deals with the study of any dynamic system. Fundamentally, it deals with 'feedback' mechanisms and is built on 'cause and effect' relations among the factors influencing the system under investigation.

It is to be noted that even though System Dynamics is grounded in Control Theory and the Modern Theory of Non-linear Dynamics, its applications may extend from managers and policy-makers to educationalists and biologists.

1.3 COMPLEX SYSTEMS AND SYSTEM DYNAMICS

The study of Complex Systems began in Physics and Chemistry initially. The study of Engineering or Business Management, Mathematics or Science, Physical Systems or Biological Systems, in terms of the *parts*, *the whole* and *the relationships* is called Complex System Analysis. In some cases, the definitions of System Dynamics and Complex Systems may overlap in the sense that both deal with the three terms just mentioned.

So, a Complex System is a combination of two or more *parts* interrelated in a systematic way to produce the *whole*, which has a typical behaviour. A car running as a result of the energy conversion through interconnected mechanisms, the disease propagation through defferent media, organizational change as a result of changing business conditions, the environmental changes, or for that matter a family, are all Complex Systems. A family falls into the category of a complex system because, it consists of *parts* which could be father, mother, and children. It is one unit as a *whole*. Also, there are different forms of *relationships* between the individual members such as father, husband, brother, etc. The theory of complex systems gives a systematic method of identifying the individual *parts*, and *relating* them to form the *whole* so that the behaviour of the entire system can be predicted or explained. The procedure could be analytical for greater depth of understanding, and some of them can be computer-based for elaborating, describing, modelling, or simulating of the complex systems. It is very important to note that in a Complex System there is always two or more possible consequences of the interaction between the parts of the system.

While tools like Critical Path Method or Program Evaluation and Review Technique can identify the critical path and provide means to review the activities to optimize the results, System Dynamics can be of assistance in studying the behaviour of the dynamic system so that one can know exactly how the system would behave under changing conditions of the controlling parameters through mathematically simulated models.

Complex Systems and System Dynamics provide a model that can inform every aspect of business management by linking research, development, conceptual design and imaginative market exploration, human resources management and overall business strategy. It deals with the entire 'Knowledge Dynamics' of the company right through need identification, creation, evaluation, selection, and implementation. Complex Systems also provide means to maximize new product delivery and adaptation to changing conditions. In the present global markets and the rapidly changing business environment, e-commerce, telecommunication and Internet, Complex Systems thinking and System Dynamics approach to simulate the environment are indispensable tools to foster creativity, maximize productivity, enhance performance, promote innovation, and increase customer satisfaction.

1.4 SYSTEM BEHAVIOUR

Any system, may it be physical or non-physical, has a typical behavioural pattern which depends upon the associated *events* and the *structure* of the system. There are two ways a model behaviour can change, one by varying the events, and two by changing the structure of the system.

Events trigger the behavioural patterns of a system. Recession, for example, is an event and the people losing their jobs, real estate rate collapse and the decline of country's economy are all the behaviours which can be observed as a result of recession. Similarly, globalization is an event and the increase in employment, reduction in cost, growth of market, increased choice of goods, economic growth of the country are the observed behaviours. But in both these examples, either the growth or decline of economy follows the structure of the system. Hence, *events* cause a *behaviour* based on the system *structure*.

The fundamental assumption in Systems Theory is that if one wishes to study the behavioural pattern of a system, the focus should be on the model *structure*. Then what exactly is this structure? In simple words, it is a set of *information interconnections* between the events that generate a system behaviour. Hence, to change the system behaviour we need to either implement changes in the events or vary the information interconnections between these events.

Let us, for example, consider the hiring and firing phenomenon in the manufacturing sector where employees are recruited and terminated based on the production requirements of the company. Based on the increased demand for the product, the production manager may hire more workforce, which would result in higher production and increased inventory (accumulation of production less shipments). This may result in higher production than required, which adds to the total cost as a result of which some employees may have to be fired. Now, the attrition may result in lower production, and thus, bringing down the inventory level. Subsequently, if the inventory falls beyond the required level the manager may not be able to meet the increase in demand if any, and hence may have to hire more employees. The ups and downs of the inventory level is the *behaviour* exhibited due to the hiring and firing of the employees which are the *events* for the observed period of time. Now the point to be observed here is that for a better control of the model behaviour (say, minimum inventory fluctuation), the *information interconnections*, i.e., the *structure* of the model needs to be robust.

1.5 APPLICATIONS OF SYSTEM DYNAMICS

There cannot be a system that is static! Whether it is a country waging a war on the other, or Buddha sitting in a state of meditation, for a student of System Dynamics, both are dynamic in nature. This is because in a war condition starting from the decision to plan for an attack, to the launching of the attack, there is dynamics involved, and in case of stagnant Buddha on meditation, the physical body may appear to be apparently static but the mind on a different plane is on a dynamic state, it may be in the fourth dimension! As Einstein rightly said, it is the 'frame of your reference' or in the context of System Dynamics the 'boundary of your system' that defines the condition of static equilibrium or the dynamic motion. So practically speaking, there is no real-life situation where System Dynamics cannot be applied. Nevertheless, from the academic point of view, following are some typical applications which are worth considering:

- Climate and environment dynamics, e.g., study of global warming phenomenon
- Energy dynamics, e.g., study of electrical energy consumption
- World dynamics, e.g., study of natural resources consumption
- Urban dynamics, e.g., study of land utilization, employment, business growth, etc.
- Business and economy dynamics, e.g., study of market growth
- Social sciences dynamics, e.g., study of dynamics of social welfare policies
- Industrial dynamics, e.g., study of inventory

- Science dynamics, e.g., study of biological systems
- Mathematics dynamics, e.g., study of statistics for historical fit
- Management dynamics, e.g., study of dynamics of decision-making
- Engineering dynamics, e.g., study of dynamics of machines
- Technology dynamics, e.g., study of dynamics of information systems
- Manufacturing dynamics, e.g., study of dynamics of supply chain of a manufacturing sector.

So, it goes without saying that System Dynamics has a wide range of application from arts and science to technology. This book lays special emphasis on multidisciplinary areas of Engineering, Technology and Management. Engineering Management is a relatively new discipline, which basically uses engineering tools and techniques in solving management problems. As such, System Dynamics in the context of this book is dealt with as an Engineering Management subject, as it uses feedback control theory in solving management related problems. The underpinning knowledge gained through the concepts given in this book may be extended into any specific area of choice where there is a scope for modelling a system in general.

The approach used in this book is to cover the salient features of System Dynamics model building in brief so that the reader will be able to acquire the fundamental knowledge on the theory of System Dynamics and can understand some case studies and projects so that it may be applied in the area of specific interest of the reader. There are many commercially used software viz. Vensim®, iThink™, STELLA™, etc. and the reader may use any one of them to build the models discussed in the subsequent chapters.

1.6 BASIC DEFINITIONS

Feedback – It is a process whereby some proportion of the output signal of a system is passed (fed back) to the input.

Causal Loop Diagram (influence diagram) – It is the pictorial representation of the 'cause and effect' relationship between the individual parameters of a system. This diagram does not indicate the stock (level) or the flow (rate) but merely indicates the influence of one parameter on the other.

Stock and Flow Diagram – It is the pictorial representation of the behaviour of the system in the form of accumulation (stock or level) and flow (rate).

Simulation Model – Mere causal loop or stock and flow diagrams do not constitute the system dynamics model. But when the parameters in these diagrams are related in the form of algebraic equations, it forms the Simulation Model.

Simulation Graphs – This is the output generated when the simulation models are fed into System Dynamics software packages. These graphs show the pattern of variation of the study variables, usually over a period of time. These graphs may be 2-dimensional or even 3-dimensional based on the requirement of the user and the software used for analysis.

Inventory – The accumulation of production less shipments, sometimes may even be referred to as stock.

1.7 SUMMARY

This chapter gives the background of System Dynamics and explains in brief the scope of the subject – System Dynamics. Some common applications of System Dynamics have been listed and some basic terms commonly used in System Dynamics have been defined.

1.8 EXERCISE

Perform an exploratory search in the Internet to:

1. List the various software packages available for solving problems in System Dynamics.
2. List 5 applications of System Dynamics other than the ones given in this chapter.
3. Go through the 'System Dynamics Society' home page in the Internet and note down the activities and the possible employment opportunities.

2

Building System Dynamics Model

Objectives

After going through this chapter you will be able to:

- Understand the Structure of System Dynamics.
- Know the various stages in the Process of System Dynamics.
- Know the various forms of Model Behaviour.

2.1 STRUCTURE OF SYSTEM DYNAMICS

The concept of structure has been explained in the preceding chapter and here the emphasis in on the constituents of the model. Any complex system may be divided into smaller subsystems and each of the subsystem can be broken down into individual components in the form of activities, events, tasks, etc. depending upon the type and nature of the system being analyzed. In System Dynamics (SD), each of these variables is related to the other in the 'causal' form. Now, by studying the causal relations of individual subsystems, the dynamic behaviour of the entire system can be predicted to the given degree of accuracy.

2.2 CAUSAL RELATIONS

If the relationship between two variables is such that change in one variable is the consequence of the change in the other variables, then a causal relation exists between the two. The independent variable becomes the 'cause' and dependent variable becomes the 'effect'. It is important to note here that a 'correlation' between two variables need not necessarily establish a 'causal relation' between the two variables.

In the study of SD, establishing the causal relation between the variables of study is the starting stage of model development. While developing a model, it is advisable to list the variables of study and all the associated variables so that they may be related causally. The variable may have positive or negative influence on each other, e.g., acceleration (cause) has positive influence on speed (effect) of a vehicle (a given amount of increase in acceleration will result in a certain amount of increase in speed, hence a positive causal relation), similarly deceleration (cause) has negative influence on speed (effect) of a vehicle. These two situations in the causal diagram may be represented as shown in Figure 2.1.

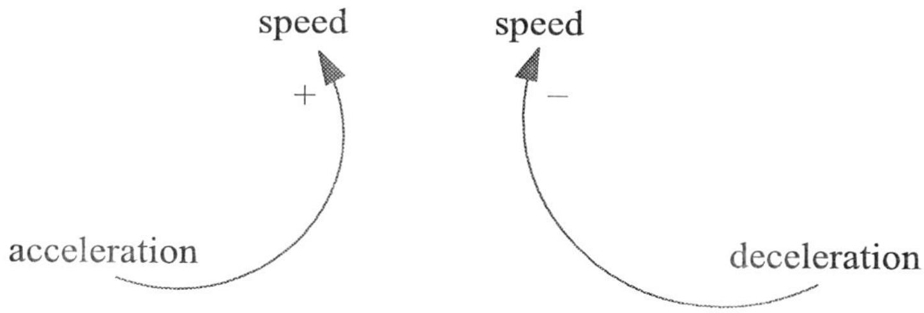

Figure 2.1: Causal Relations.

Hence, to start with, it is important to list all the variables of study and link them in the form of cause and effect so that the nature of relationship between all the variables is clear. It is possible that an effect can be the cause for another effect and the chain may continue. In other words, an independent variable may cause change in a dependent variable, which may in turn cause a change in the third variable, and so on.

2.3 FEEDBACK SYSTEMS

In terms of System Dynamics the systems in nature can be classified into 'open' or 'closed' systems. In simple terms, if there is no feedback, it becomes an 'open system' and if the feedback exists, it becomes a 'closed system'. Going a step further, in both the systems, output of the system responds to the input, but in case of open system the output has no influence on the input, whereas, in a closed system the output does influence the input.

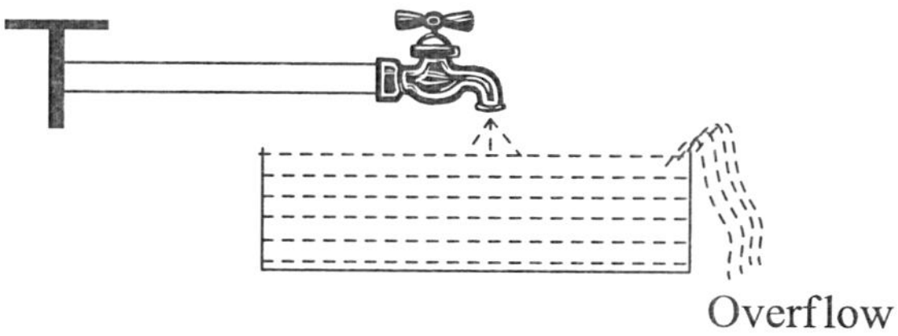

Figure 2.2: Open System.

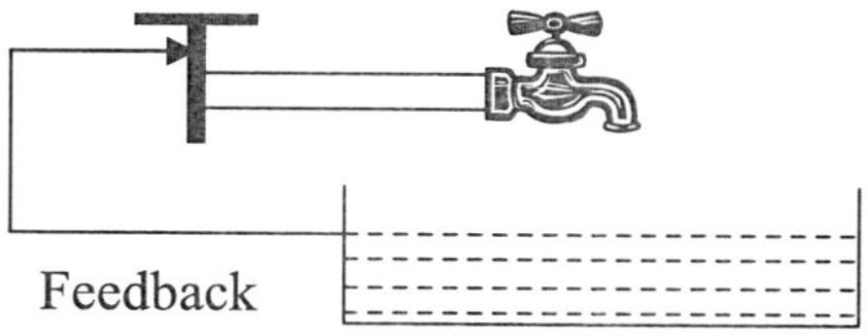

Figure 2.3: Closed System.

Consider the simple case of a tap and a tank where the input is the water flow into the tank and the output is the rise in water level. Here, the tap regulates the flow of water and the tank contains a certain amount of water. Figure 2.2 represents an open system because once the tap is open the water level continuously increases and it may even overflow the tank. On the other hand, Figure 2.3 represents a closed system because it has a feedback on the water level and as the predetermined level is reached the feedback mechanism automatically closes the tap. This can be even a continuous operation if the feedback sends a signal even at the minimum level of water. In such a case when the water level reaches a maximum level the feedback may close the tap and when the water level comes down to a preset minimum level the feedback may again switch on the tap. Hence, it is clear that in the first case the system input (flow of water into the tank) is not controlled by the output (rise in water level in the tank), and hence it is an open system, whereas, in the second case the input is controlled by the output of the system and it is a closed system.

In system dynamics the feedback systems are dealt with under the heading of Cybernetics. It is a study of the system with feedback structure and is applicable to both physical and social systems. Cybernetics is basically interdisciplinary as it connects various disciplines such as: control systems, electrical engineering, mechanical engineering, biological systems, anthropology, and so on.

2.4 REINFORCING AND BALANCING LOOPS

During the formation of causal diagrams, two types of loops may be formed, viz., Reinforcing or Positive loop and Balanced or Negative loop. The identification of this nature of loop can make some key revelations regarding the model behaviour, and hence, after drawing the causal loop diagram it would be very useful to identify the possible loops in the structure.

2.4.1 Reinforcing or Positive Loop

In case of positive loops an action triggers a result that generates more of the action and subsequently more of the result, and hence, the process is 'reinforcing'. The fundamental characteristic of the positive loop is that it disturbs the stability of the system thus leading to a growth pattern or a decay pattern.

Consider the case of increase in quality of a product, which results in more sales, which in turn increases the revenue that can be used to further increase the quality and this constitutes a reinforcing cycle. The positive loop and the growth of Sales and Revenue with the increase in quality are shown in Figures 2.4 and 2.5 respectively.

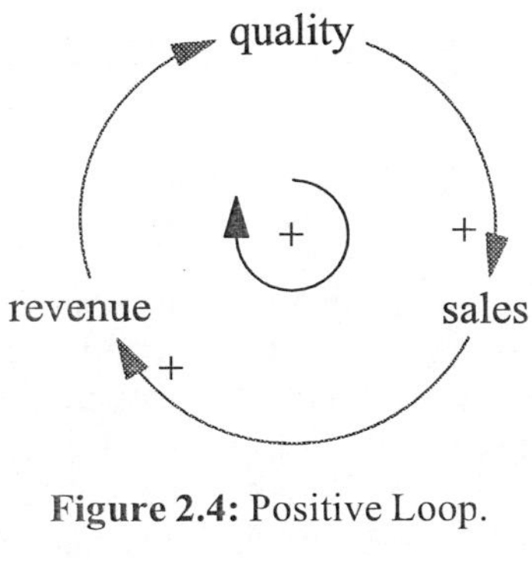

Figure 2.4: Positive Loop.

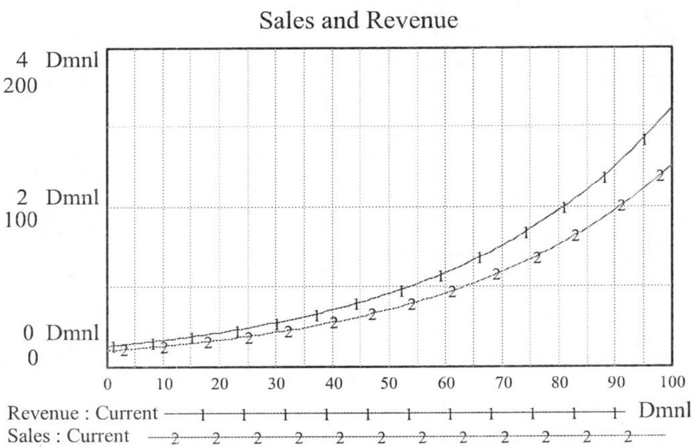

Figure 2.5: Growth Pattern.

2.4.2 Balanced or Negative Loop

Let us consider a case of increase in number of disease, which decreases population (due to increase in death rate), and increase in population increases the number of diseases subsequently (due to several reasons such as pollution, disease propagation, number of susceptible patients, etc.). The causal loop and the corresponding decreasing pattern are shown in Figures 2.6 and 2.7

respectively. It has to be noted that even though with the increase in the number of diseases, population decreases but the simulation is for the total number of diseased persons for a given population size, and hence, both are showing the decreasing trend.

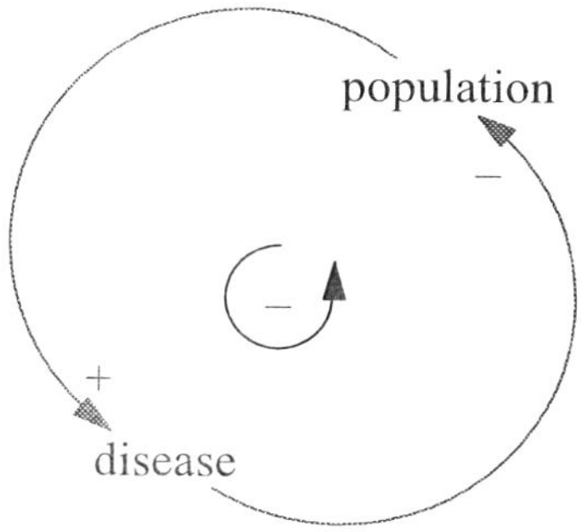

Figure 2.6: Negative Loop.

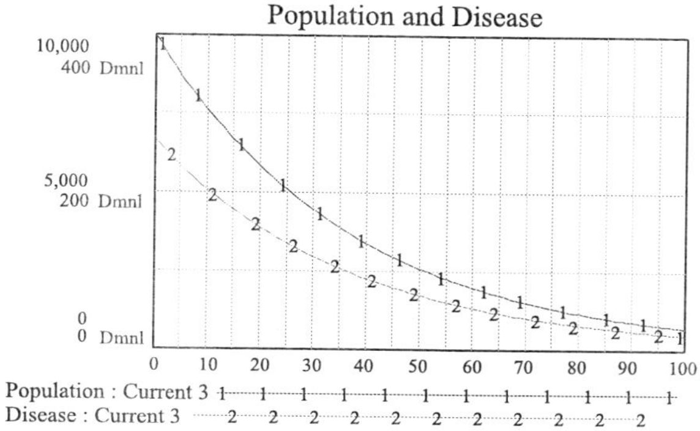

Figure 2.7: Decaying Pattern.

2.4.3 Combination of Positive and Negative Loops

Quite often the loops mentioned above act in conjunction with each other. This is better explained with the Population Growth Model (Figure 2.8). Population enclosed inside a rectangle indicates that it is level or stock (explained in the Section 2.5). Usually, as a convention, variable is written in lower case and level is capitalized as shown in the causal loop.

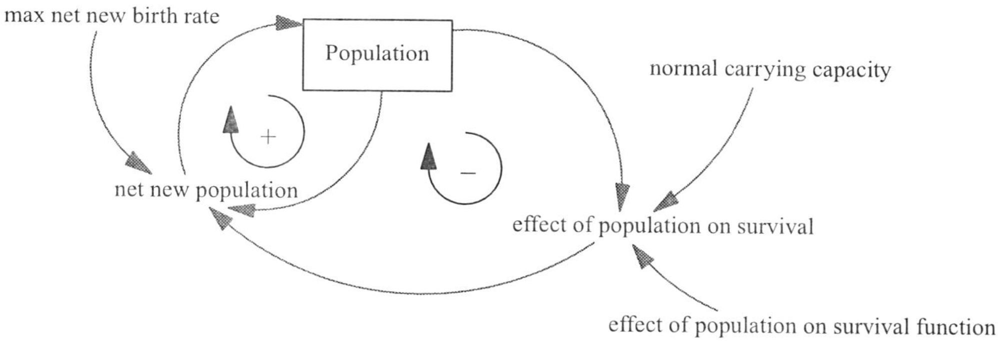

Figure 2.8: Combination of Positive and Negative Loops.

Population growth is a combination of positive and negative loops. The positive loop comprises net new population (depends upon existing *population, max net new birth rate, and effect of population on survival)* and *population*. The negative loop comprises *effect of population on survival* (depends on *normal carrying capacity of the system and effect of population on survival function*), *net new population* and *population* (Note: *Effect of population on survival function* indicates the pattern of variation which could be given in the form of a curve or the data points in the software used). It can be observed that the combined variation of population (Figure 2.11) is based on the influence of the positive and negative loops which happens to be a S-curve even though the variation of net new population (Figure 2.9) and the *effect of population on survival* (Figure 2.10) follow a different pattern altogether.

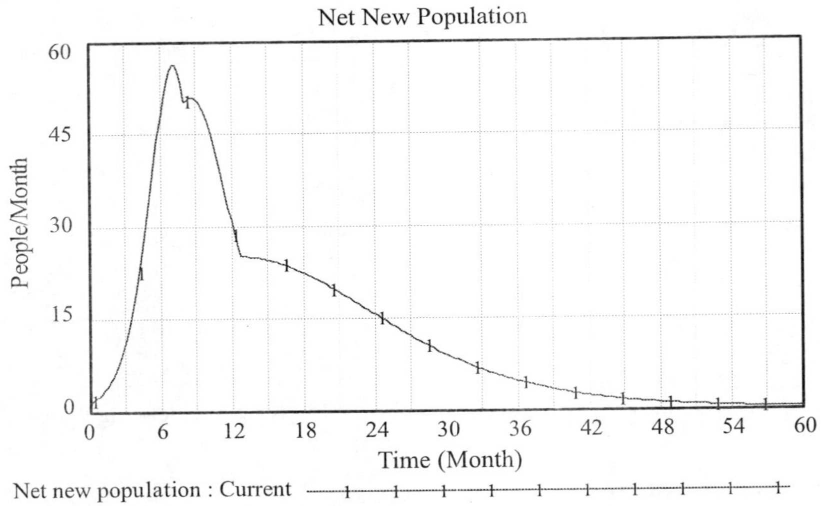

Figure 2.9: Variation of Net New Population.

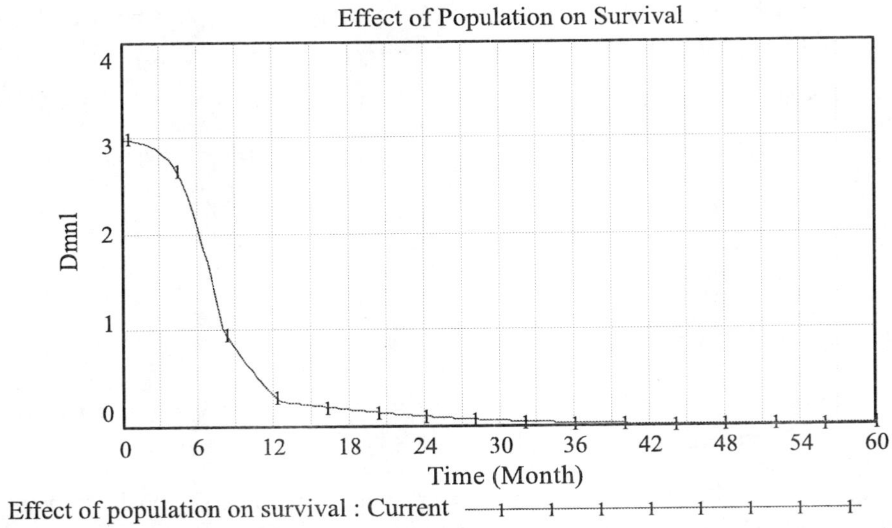

Figure 2.10: Effect of Population on Survival.

2.5 STOCK AND FLOW DIAGRAMS

A typical SD model involves two elements, i.e., rate and level. Rate (or flow) is the control and level (or stock) as the name implies is the accumulation point, and hence, the name 'stock and flow' diagram. The importance of the model structure has been explained before and in fact the

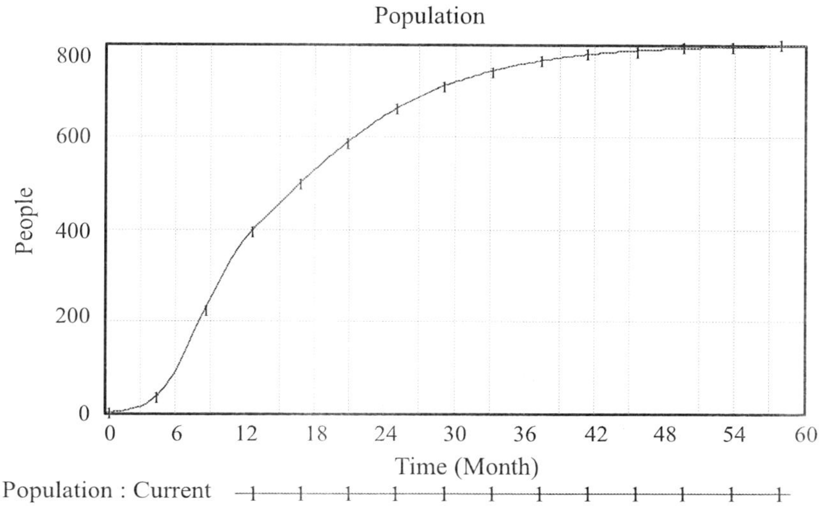

Figure 2.11: Variation of Population.

stock and flow diagrams better represent the structure of the model based on the governing equations, in comparison to the causal loop diagrams which primarily depict the causal relations.

Stocks or levels are the points where the behaviour of the system is recorded and flows or rates cause the stocks to change. While Causal Loop Diagram (CLD) gives and understanding of the causal relations between all the associated variables, the stock and flow diagrams actually constitute the first step in the building of the SD model.

2.6 SYSTEM DYNAMICS PROCESS

The System Dynamics process is basically an iterative process. Systems are usually controlled by the variables and the nature and number of variables, which govern the system, are not easy to predict, and also, new variables keep adding as the system keeps on developing. Following are the important stages of System Dynamics process:

2.6.1 Problem Statement

'A well-defined problem is half solved', goes the adage. In the first phase of the System Dynamics process, the purpose of the problem is to be clearly identified. The understanding of the problem will help the process of model development. In the problem statement stage, the expected result from the System Dynamics model will be clear in the mind of the modeller.

2.6.2 Variable Identification

Based on the problem statement, certain key variables which significantly influence the behaviour of the system need to be identified. Further, the sub-variables may also be identified such that a list of all the variables, as decided by the modeller, is recorded for further analysis. The list may keep growing as greater insights are drawn, but to start with, the most important variables have to be identified at this stage.

2.6.3 Reference Modes

The reference modes are the system characteristics. They refer to the general trend that represents the characteristic behaviour of the system. For example, the sudden oscillations that may take place in the graph, which results due to the hiring and firing that is taking place in the labour market may be a reference mode of interest. Reference modes may be the past behaviour that has been observed or the future behaviour that is anticipated.

2.6.4 Reality Check

To check whether the model agrees with certain key issues which are obvious, the Reality Check is performed. In a given model, some interrelations are inevitable. Nevertheless, the influence of one variable on another variable may also be of interest in this stage so as to get a better understanding of the system. Reality checks are based on the earlier experience of the modeller on the problem being tackled, e.g., with the demand, the supply must also increase.

2.6.5 Dynamic Hypothesis

Hypothesis is a statement that can be proved right or wrong, after a thorough research based investigation. In System Dynamics, a dynamic hypothesis may be proposed in the form of a statement, causal loop diagram, or stock and flow diagram. It basically formulates the model for conducting simulation by deciding on the variables of study. Refinement and revision can be made on this dynamic hypothesis, as and when required, because no model can be perfect in all respects.

2.6.6 Simulation

Simulation is the final stage of Dynamic Hypothesis. Here, all the variables are related using mathematical equations. Simulation generates patterns through which the behaviour of the system can be studied. In today's world, simulation plays a vital role in predicting the behaviour of complex dynamic systems such as: spread of disease, increase in population, biological systems, pollution control, environmental protection, industrial dynamics, traffic control, urban dynamics, waste disposal, manpower planning, technology management, etc.

2.7 MODEL BEHAVIOUR

The structure of the model governs its behaviour. Basically, there are four types of model behavior, which are observed frequently in the System Dynamics models. At this stage, it is important to understand the diagrams, however, model building using the software packages (Vensim®, iThink™, STELLA™, MATLAB Simulink, etc.) can be attempted for the exercise given in the next chapter and the subsequent case studies.

2.7.1 Exponential Growth

Consider an organization that keeps making profit and grows in size continuously. The recruitment rate (number of employees/year) is fixed, say, based on the experience from previous years, and recruitment takes place yearly on a continuous basis. This forms a positive loop (reinforcing).

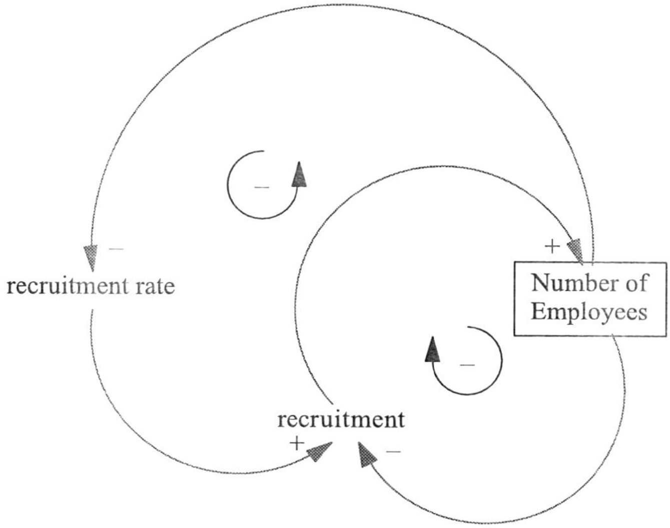

Figure 2.12: The Causal Loop Diagram (Recruitment).

It would be interesting to note how the number of employees would grow over a period of time. The causal diagram of this situation is shown in Figure 2.12. (It may be noted that a variable is indicated in small letters and the 'level' is capitalized.) Suppose if the employment rate is 1% of the total employees per year, the growth of employees would be exponential as indicated in Figure 2.13. There would be little or no significant change in the beginning years, but at the end there will be a rapid increase in the number of employees. The stock and flow diagram for the above situation is shown in Figure 2.14.

Figure 2.13: Growth Curve of Employees.

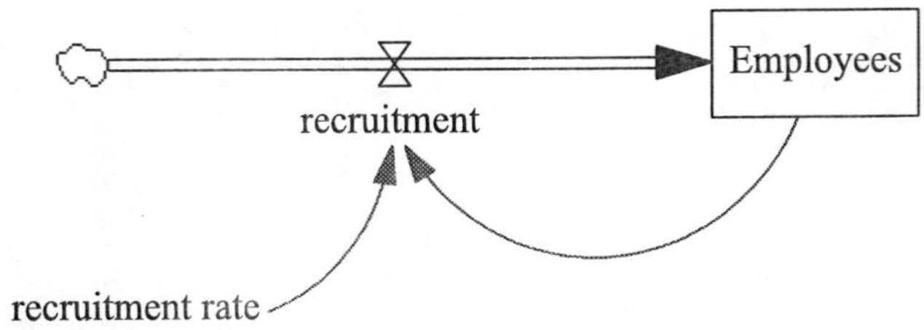

Figure 2.14: Stock and Flow Diagram (Recruitment).

2.7.2 Exponential Decay

Consider the case of a company not doing well and the employees decide to quit as soon as they get an alternative. Suppose the company had 1000 employees to start with an average worker remaining for about 10 years. Now, this is a case of negative feedback (balancing) (Figure 2.15).

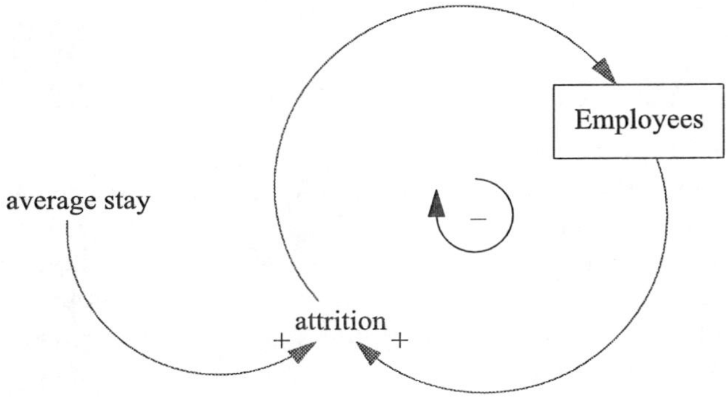

Figure 2.15: Causal Loop (Attrition of Employees).

If we plot the number of employees over a period of time, it can be observed that exponential decay takes place with a sudden fall in the number of employees followed by a gradual decrease (Figure 2.16).

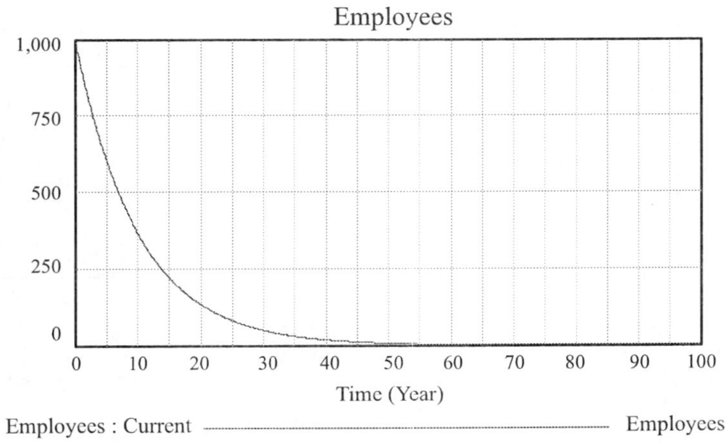

Figure 2.16: Exponential Decay (Employee Attrition).

The corresponding stock and flow diagram is as shown in Figure 2.17.

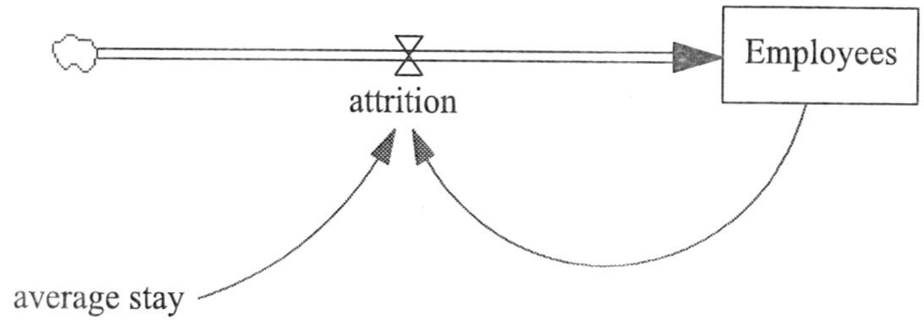

Figure 2.17: Stock and Flow Diagram (Employee Attrition).

2.7.3 S-Shaped Curve

If there happens to be two loops in a Dynamic Hypothesis, then the resulting response could be S-shaped curve. Let us consider a case of production, where the 'net new production' is causally linked to the 'effect of density on production' as shown (Figure 2.18).

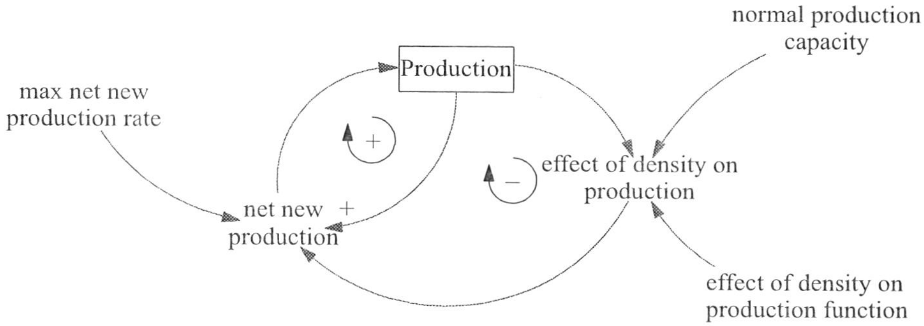

Figure 2.18: Causal Diagram (Production Model).

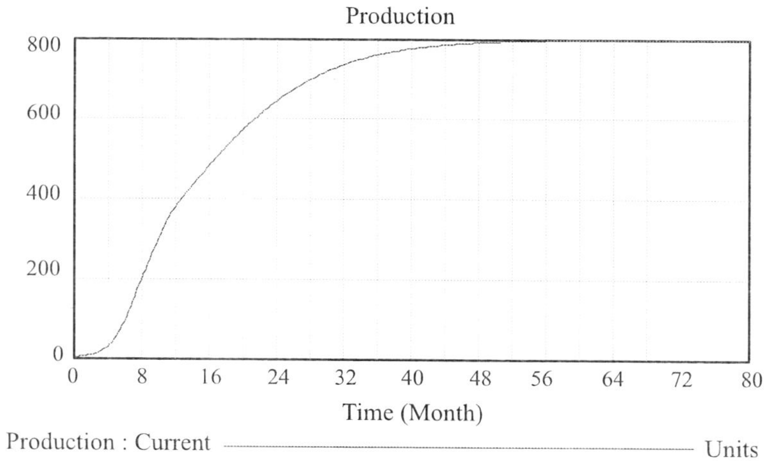

Figure 2.19: Variation of Production over Time Period.

Initially, increase in production starts like an exponential growth with the positive loop dominating (Figure 2.19). But after some time, the negative loop starts dominating such that the number of units will never exceed a specified number. This is an example of non-linear variation.

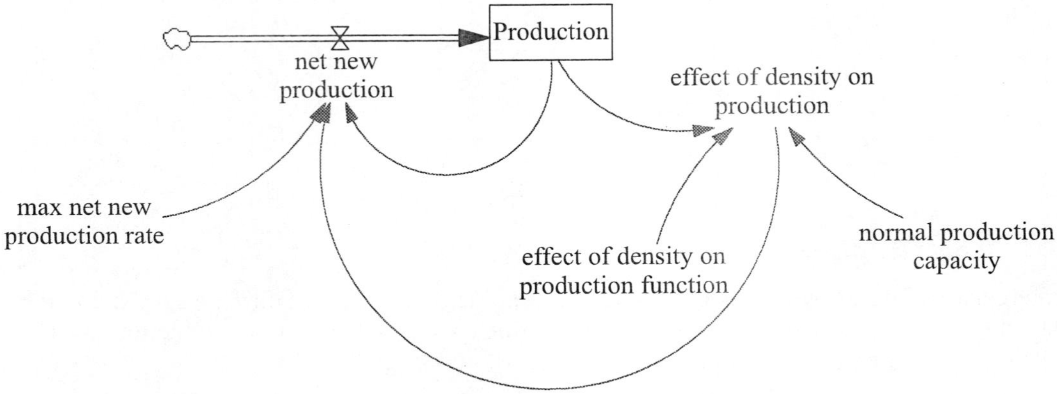

Figure 2.20: Stock and Flow Diagram (Production Model).

A typical stock and flow diagram for the production model is given in Figure 2.20.

2.7.4 Oscillation

Consider a simple spring mass system under forced vibration (Figure 2.21).

$x = v.t;$ $v = a.t;$ $a = F/m\ F = -kx$

Note: Integration of Velocity (v) gives Displacement (x)

Integration of Acceleration (a) gives Velocity
F = Force (Newton); m = Mass (kg); and k = Spring constant (Newton/metre).

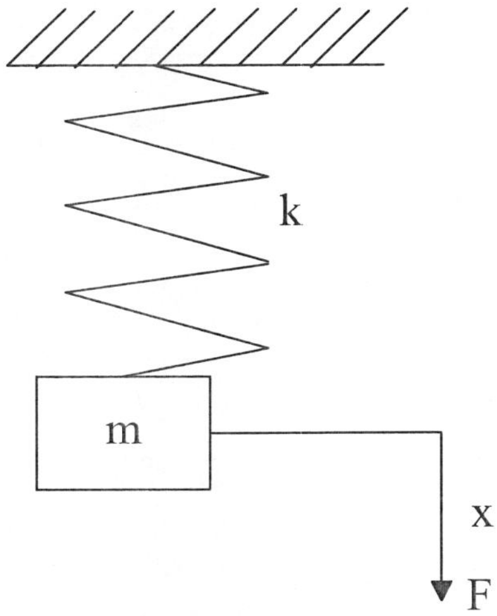

Figure 2.21: Spring Mass System.

The entire system may be represented in the form of causal loop as shown in Figure 2.22.

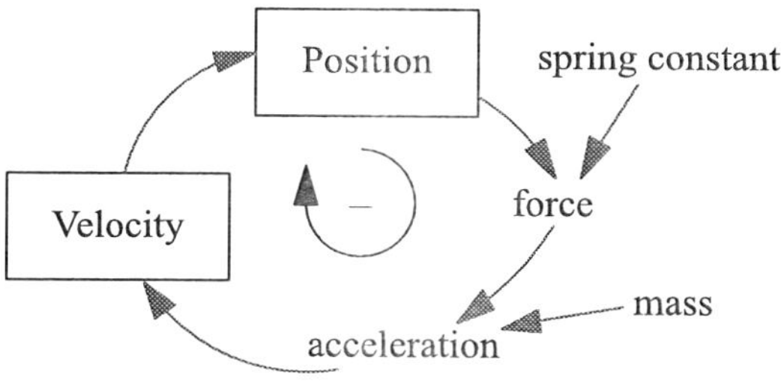

Figure 2.22: Causal Loop of Spring Mass System.

If one is interested in recording the displacement (position) at any instant of time, the graph would be as follows (Figure 2.23).

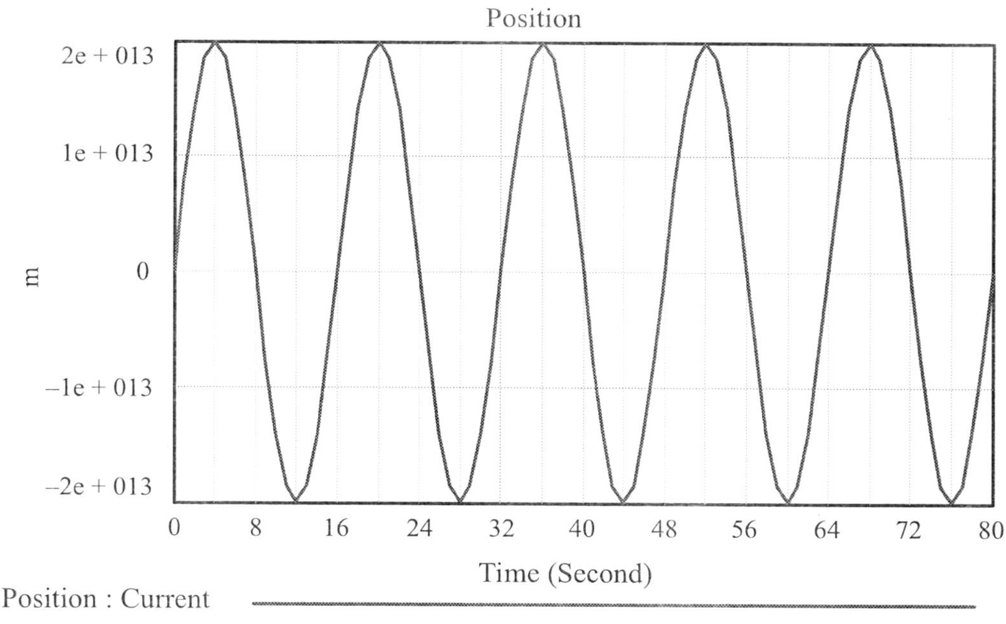

Figure 2.23: Simulation of Position versus Time.

In addition to the above four basic behaviour patterns typically observed in System Dynamics models, there are also other behaviours two of which are:

1. S-shaped growth with overshoot.
2. Overshoot and collapse (Figure 2.24).

These diagrams will be shown in specific case studies dealt with in the subsequent chapters.

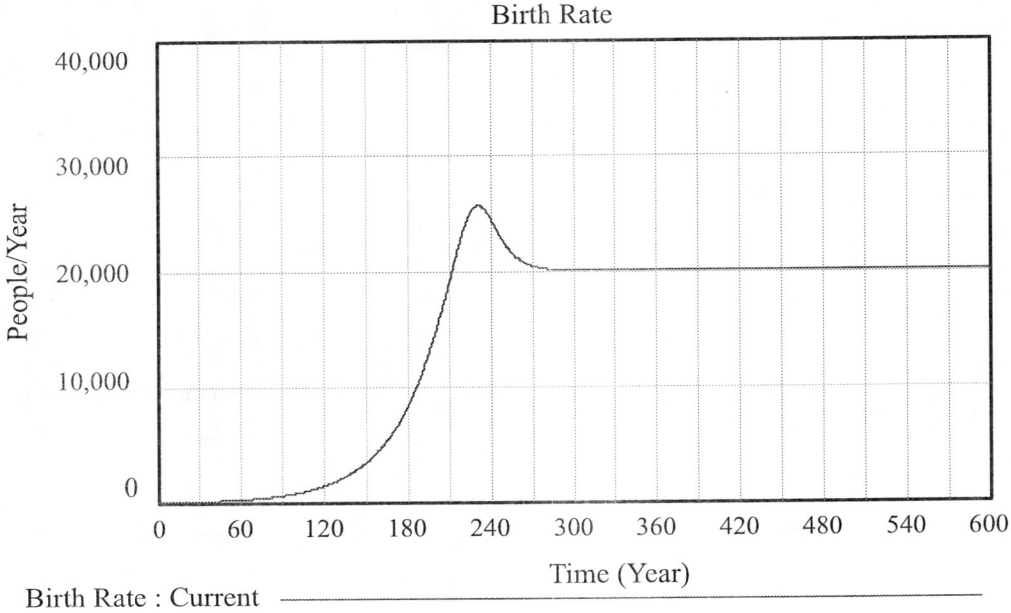

Figure 2.24: Overshoot and Collapse.

2.8 SUMMARY

This chapter deals with the Structure of System Dynamics and explains the various stages in System Dynamics Process. It also explains the Model Behaviour commonly observed in the System Dynamics problems.

2.9 EXERCISE

Can you think of a simple problem in Management, say *Hiring of employees* and how it would influence the *Number of employees* by considering *attrition* into account and draw the causal loop diagram and stock and flow diagram?

3

Human Resource Planning Exercise

Objectives

After going through this chapter you will be able to:

- Define a simple HR Planning problem.
- Identify a problem situation in HR planning and apply System Dynamics methodology to solve the same.

3.1 PROBLEM STATEMENT

Company XYZ manufacturing table fans requires HR planning for future years. Even though individual components are outsourced, the final assembly and quality check takes place in accordance to the company standards.

The production manager has based his production on current sales, as a result of which, the production has resulted in higher or lower values than necessary. This has a direct bearing on the inventory. So, the manpower requirement is to be adjusted to suit to the requirement of optimum *production* and *inventory*.

Based on past experience, the HR planning department has arrived at a *target inventory* and *target manpower*. *Target inventory* is decided based on the sales and inventory coverage. Also, target manpower is based on target production and productivity. Further, based on the inventory, the inventory correction is applied, which also depends on the target inventory. As the entire operation is time based, two time factors, namely, time to adjust manpower and time to correct inventory are also fixed.

The HR planning manager wants to build a System Dynamics model and produce simulations to study:

1. The effect of change in time to correct inventory on manpower.
2. The effect of time to adjust manpower on manpower.

Note: It is a usual practice to write a parameter (e.g., target inventory) in a system dynamics model in small letters and the stock variables capitalized (e.g., Inventory) while representing them in the stock and flow diagrams.

3.2 SOLUTION

3.2.1 Causal Tree

The causal tree diagrams for Inventory and Manpower are shown in Figure 3.1.

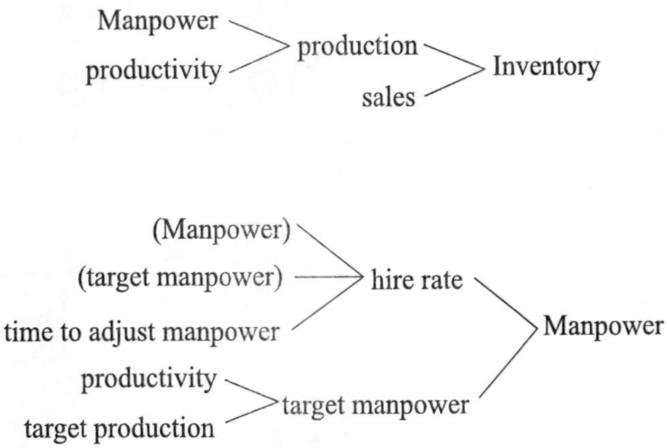

Figure 3.1: Causal Tree Diagrams of Inventory and Manpower.

Developing the causal diagrams (Figure 3.2) is a very important stage in the System Dynamics model development. Following guidelines may be useful in building causal diagrams:

- List all the parameters that control the HR planning.
- Develop causal relations between the individual parameters.
- Indicate whether the parameter has a 'positive' or a 'negative' influence on the associated parameter.
- Finally, check whether the parameters selected are sufficient to describe the model in full. In case some variables are left out, incorporate them.

The model given in Figure 3.2 is not a complete model. The purpose here is to show the causal relations among the HR planning variables in the most fundamental form. The emphasis is on relating the variables 'Production' and 'Manpower'.

3.2.2 Governing Equations

Once the causal relations are developed, the variables can be linked in the form of equations. For example, the variables shown in Figure 3.3 and their governing equations are related in the form given below. The units of these variables are indicated in the parentheses. It is important to note that the units of different variables in a given equation should match to run simulation.

Production = Manpower × productivity; (units/month)

Inventory = initial value (a constant, 200 say) + ∫ (production-sales); (units/month)

Manpower = initial value (target manpower) + ∫ (net hire rate); (person)

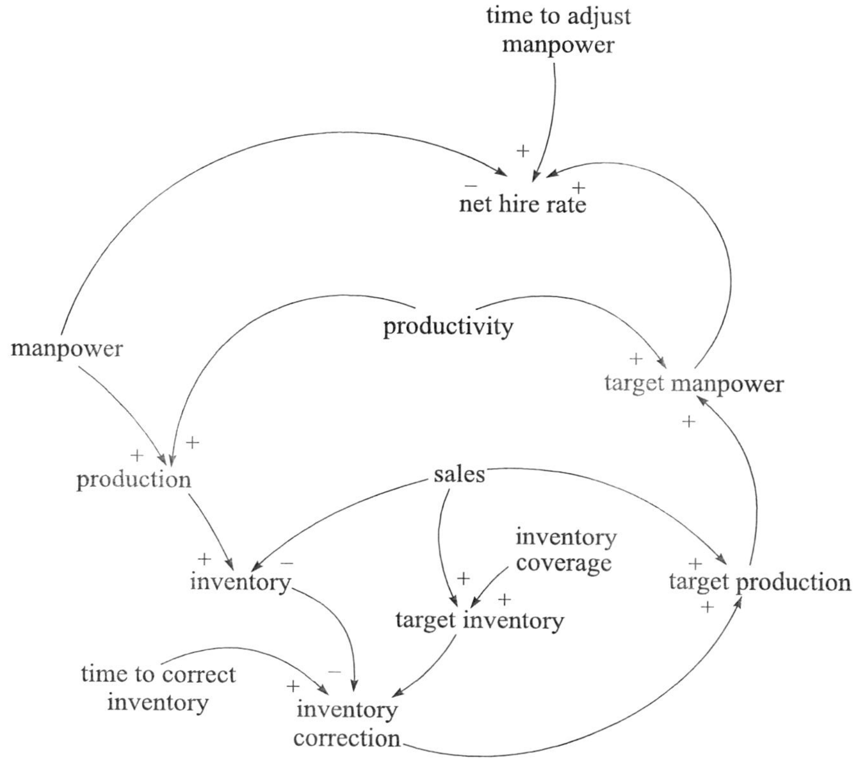

Figure 3.2: Causal Diagram of HRP Model.

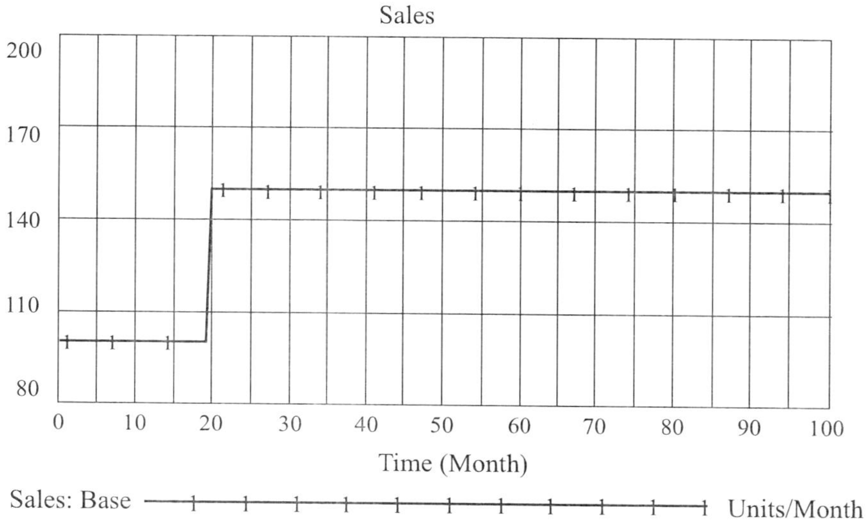

Figure 3.3: Increase in Sales as a Step Function.

Sales = 100 + STEP(50, 20) (This is a step function where the value is a constant (say, 100) up to time value of say, 20 months, and then rises to 150 (figure 15));
(units/month)

Productivity = a constant (Take as 1) (units/month/person);

Time to adjust manpower = a constant, say 1 month; (month)

Net hire rate = (target manpower – manpower)/time to adjust manpower; (person/month)

Target manpower = target production/productivity; (person)

Target production = sales × inventory correction; (units/month)

Target inventory = sales × inventory coverage; (units)

Inventory correction = (target inventory – inventory)/time to correct inventory; (units/month)

Time to correct inventory = a constant, say 1 month; (month)

Inventory coverage = a constant, say 3 months; (month).

3.2.3 Stock and Flow Diagrams

Even though the causal diagram is sufficient to produce the simulations, Stock and Flow diagrams distinguish the parameters into the controlling 'flow' acting as regulators, e.g., production rate, and 'stock' where accumulation takes place, e.g., manpower.

The above causal diagram may be represented in the form of stock and flow diagram as shown in Figure 3.4.

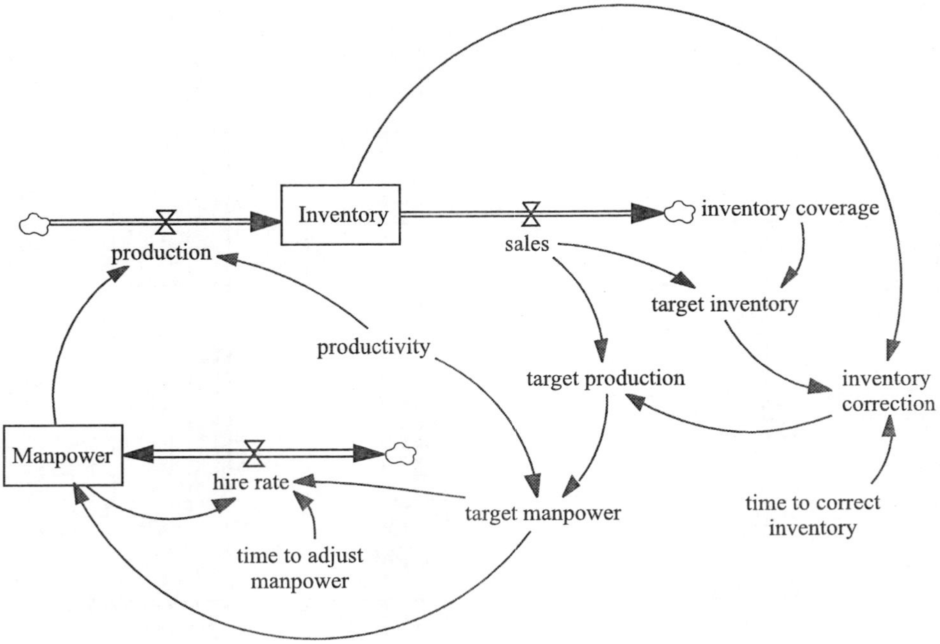

Figure 3.4: The Stock and Flow Diagram of HRP.

3.2.4 Simulation

Having broken down the complex HR planning problem into individual subsystems, and related them in the causal form, and then, having developed the stock and flow diagrams based on the governing equations, the model is now ready for simulation.

In this example, we are interested in studying the effect of two 'time-based parameters' on 'manpower requirement':

1. Time to correct inventory
2. Time to adjust manpower.

In a dynamics analysis which involves two parameters whose influence is to be studied, one of them has to be kept constant and the other varied. So, to start with, let us keep the *time to adjust the manpower* as a constant, say 1 month, and vary the *time to correct inventory*.

Case 1: Influence of Time to Correct Inventory

Following is the simulation result for time to correct inventory values as 1, 2, 3 and 4 months (Figure 3.5). The simulation shows the influence of time to correct inventory on the manpower requirement. It can be observed that the fluctuations settle down over a period of time. It can also be observed that after a period of 30 months, the manpower requirement uniformly settles down to 150 and remains constant, irrespective of the change in the time to correct inventory, and hence, indicates that the system stabilizes.

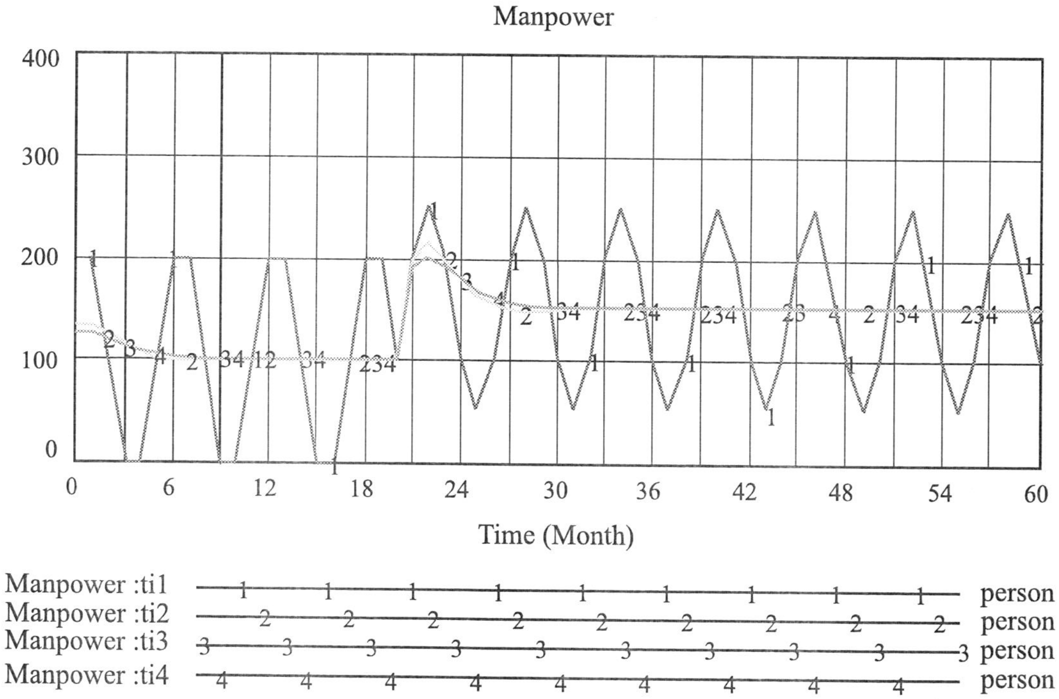

Figure 3.5: Influence of Time to Correct Inventory on Manpower Requirement.

Case 2: Influence of Time to Adjust Manpower

The time to adjust manpower is varied from 1 to 4 months keeping the time to correct inventory a constant. Figure 3.6 shows the simulation result. It can be observed that as the time to adjust manpower is increased, the fluctuations in manpower requirement die out eventually, and after about 48 months the manpower requirement stabilizes around the mean value of about 150.

It has to be noted that this is not an all inclusive model but a simple model developed to explain the steps in building the System Dynamics Model and conducting simulations. Further, it has to be noted that in the above two cases, the numbers such as the manpower or months are fictitious values, but what is important is, the indication of the trend and the pattern of variation that occurs over a period of time. Both the graphs indicate that the system experiences a major disturbance on the 20th month (due to increased sales) but over a period of time it stabilizes to meet the specifications as defined through the governing equations. Hence, the simulation result could be used only to support an argument in making Policy Implementation decisions in HR planning and can never be the only source to arrive at a conclusion. In other words, the simulation results can only substantiate intuition, hunch, experience based decisions, etc. of a manager, but never be the only criterion for decision-making.

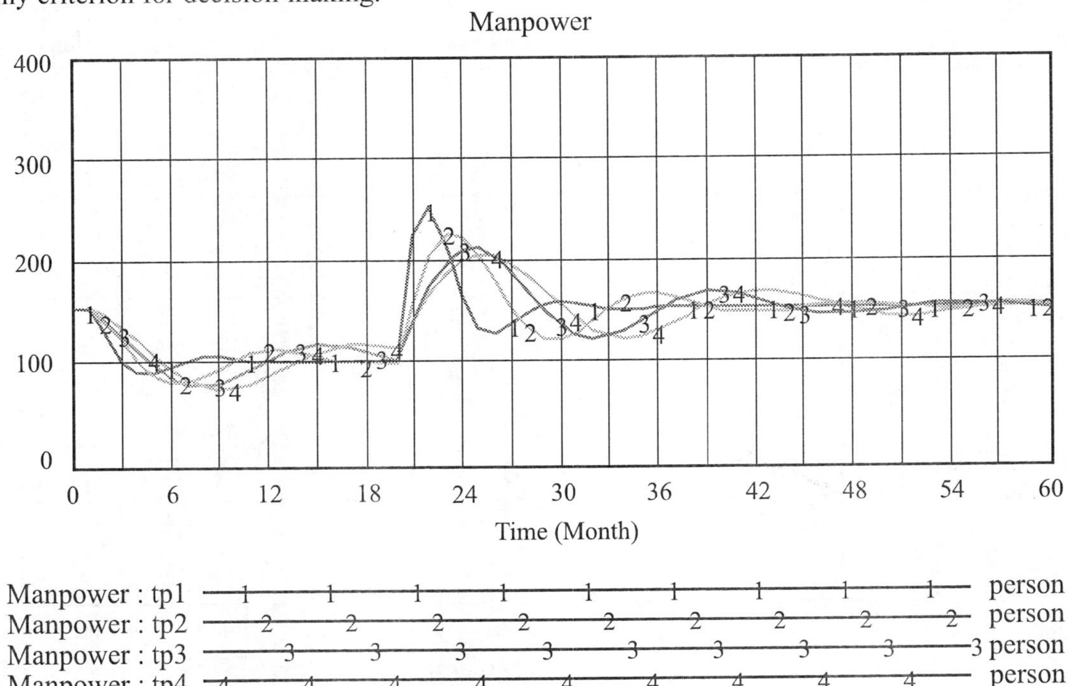

Figure 3.6: Influence of Time to Adjust Manpower on Manpower Requirement.

3.3 SUMMARY

In this chapter a simple HR planning exercise is taken as a case study and the System Dynamics process has been explained. It clearly shows how to list the associated parameters and build the causal loop diagram and the stock and flow diagram starting with the causal tree diagrams. It also explains how each of the parameter is algebraically defined so as to build the System Dynamics model. This chapter also explains how to interpret the simulation results that could be obtained from any of the standard software packages. The reader is encouraged to use the educational/trial version of any of the commercially available software packages listed in the preceding chapters to build the model and run simulation. The tutorials provided in these simulation packages will clearly give step-by-step procedure to build the models, which is beyond the scope of this book. However, the reader has to use any of the commercially available software packages if he/she intends to become a professional System Dynamics Modeller or a System Dynamics researcher.

3.4 EXERCISE

For the same model, produce the simulation graph to show the influence of the two time factors on 'Inventory'.

4

Project Management in New Product Development

Objectives

After going through this chapter you will be able to:

- Understand the meaning of 'Scope of a Project' and 'New Product Development'.
- Identify the problem associated with the Project Management in New Product Development.
- Identify the parameters which influence the project dynamics.
- Construct the causal loop diagram and stock and flow diagram for a given problem situation.
- Appreciate the purpose and use of conducting simulations.
- Interpret the simulation results and draw implications in the context of the case study under consideration.

4.1 DEFINITIONS

Scope of a project

The scope of a project is defined as the work to be completed for product delivery with the specified features and functions (PMBOK, 2000).

New Product Development (NPD)

The Product Development and Management Association (PDMA) defines NPD as: "a disciplined and defined set of tasks and steps that describe the normal means by which a company repetitively converts embryonic ideas into saleable products and services" (Belliveau et al., 2002: 455).

4.2 THE PROBLEM STATEMENT

The Project Manager of Company ABC has been advised by the CEO to study the system parameters that influence the 'change in scope' in one of their innovative products. He has narrowed down to the study of change in scope, as the other critical success factors such as cost, quality and time are also influenced by the change in scope. A lot of money is being invested on employee training and even retraining, due to the change in scope and a decision has to be made whether the investment made justifies the cost on a long-term perspective.

During their operations, it has been observed that 'schedule gap' (project completion date overshooting deadline) is a major problem they encounter from time to time. Their most common response has been to increase the headcount as soon as they anticipate a schedule gap. But their project manager, in the present context through his experience, observed that competence loss, employee productivity, and schedule pressure related aspects resulting from the 'change in scope' were the major dynamic risk factors, which were going unnoticed. So, he decides to build a dynamic model that would help in the better understanding of the project from time to time through which the solution to the problem posed by the CEO could also be addressed.

4.3 OBJECTIVES

With the above problem statement in mind, following are the specific objectives to be accomplished:

- Identifying and relating the variables involved in a typical project in which there is a change in scope.
- Developing causal loop diagram
- Constructing the stock and flow diagram
- Formulating the governing equations
- Modelling and simulation
- Drawing inferences for making policy implementations.

4.4 THE VARIABLES

To start with, a thorough study has to be undertaken to identify all the variables that may have influence on the project dynamics of NPD. Let us first list all the parameters randomly, without trying to relate them. Following are the parameters which come into our mind when we speak of Project Dynamics in NPD:

- Business growth
- Customer satisfaction
- Work completed
- Work quality
- Project success
- New projects
- Base work to be done
- Desired competence pool level
- Competence gap
- Forecast competence fail rate
- Recruitment rate
- Training rate
- Training completion rate
- Actual competence pool level
- Fatigue
- Schedule pressure
- Employee compensation
- Finance available
- R&D funding
- Revenue
- Service rendered
- Cost and performance
- Time available
- Apparent work completion rate
- Time remaining to complete work
- Undiscovered rework
- Error generation rate
- Change in scope of project
- Desired knowledge in process
- Discrepancy in knowledge in process

The above may not be an all-encompassing list. But for this project, let us consider these variables to start with, as one gets greater insight into the problem there is always a provision to add new variables.

4.5 CAUSALITIES

Let us now try to causally relate these variables in the form of loops. Any variable may either have no influence, or have a positive or negative influence on another variable. But by and large, everything in nature is dependent on everything else, directly or indirectly. For instance, consider schedule pressure as a variable. By pure logic, we can associate the variables as shown in Table 4.1.

Table 4.1: Schedule Pressure Influence

Schedule Pressure			
Influenced by		*Influence on*	
Positive	*Negative*	*Positive*	*Negative*
– Time available – Time remaining for work to complete	– Actual competence pool level	– Fatigue – Ripple effect (rise and fall) – Error generation rate	– Apparent work completion rate

The above relation may be represented in the form of causal loop diagram as shown in Figure 4.1.

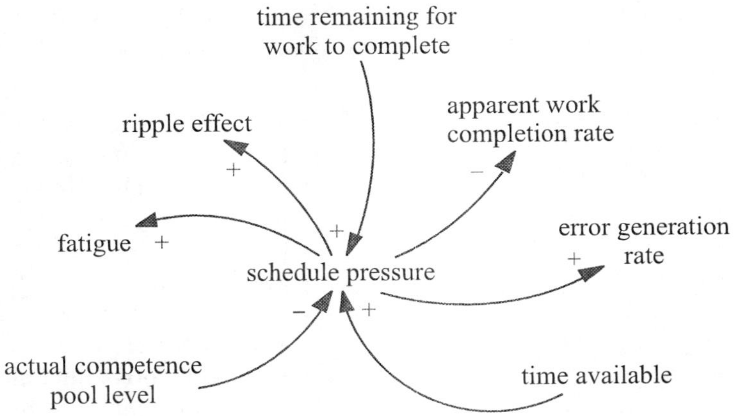

Figure 4.1: Causality of Schedule Pressure.

In the same way the causal loop diagram for the entire system of project dynamics under the NPD may be developed and is given in Figure 4.2.

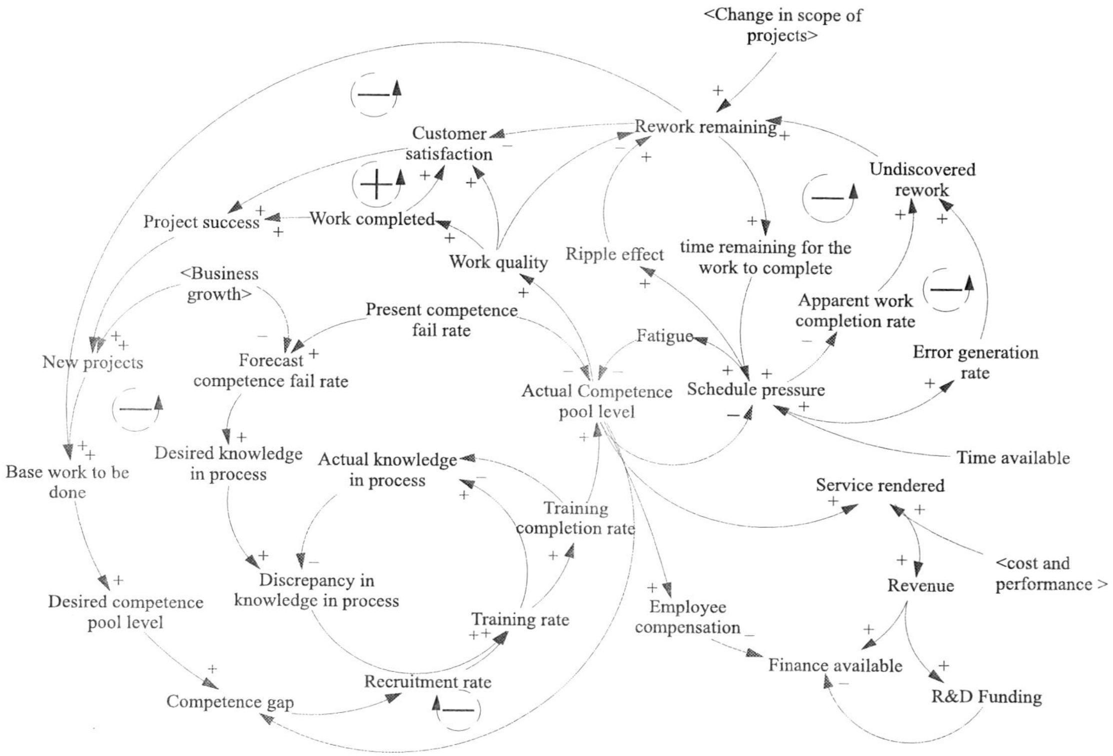

Figure 4.2: Causal Diagram for the Project Dynamics in NPD.

It can be observed that the entire causal diagram will result in five negative loops and one positive loop (refer Chapter 2 for positive and negative loops).

The causal diagram basically indicates the relationship between factors such as change in scope, rework, workforce, etc. Ideally speaking, the project should start and finish at the right time as expected or determined before the start of the project. The deviation from this occurs in practical situations due to several reasons, which ultimately contribute to the discrepancy in workforce. This calls for the hiring of additional Product Development Staffs (PDSs). Further, the PDSs are trained to acquire the required level of competency. Rework is also an unavoidable phenomenon which may be due to high quality standards, human errors, work obsolescence, machine wear-out, etc. It is important to note that some of the prime contributing factors to change in scope are: change in technology, change in market conditions, ever growing customer expectations, competition, etc. These are some of the factors to be considered by a System Dynamics modeller dealing with the Project dynamics in NPD.

4.6 STOCK AND FLOW DIAGRAMS

The modelling of the Project Dynamics of NPD is carried out in three sectors:

1. Product development workflow sector

2. Competent product development staffs sector
3. Product development salary and revenue sector.

1. *Product development workflow sector*

The desired workflow is given by the formula:

Desired workflow = (initial work)/(schedule time).

The initial work is the input to the stock (also called level), i.e., work to be done, and is controlled by the work rate (Figure 4.3). The work to be done is then stocked as work in progress, the *workflow rate* being determined by a number of variables such as productivity of workforce, workflow under normal condition, etc. Further, the work in progress is moved to stock in work finished, provided it is to the required standards as controlled by the acceptance rate. In the level work in progress, rejected work has to be reworked, which introduce two more rate controls as shown. The work finished is subject to obsolescence and is subjected to rework based on the control obsolescence rate.

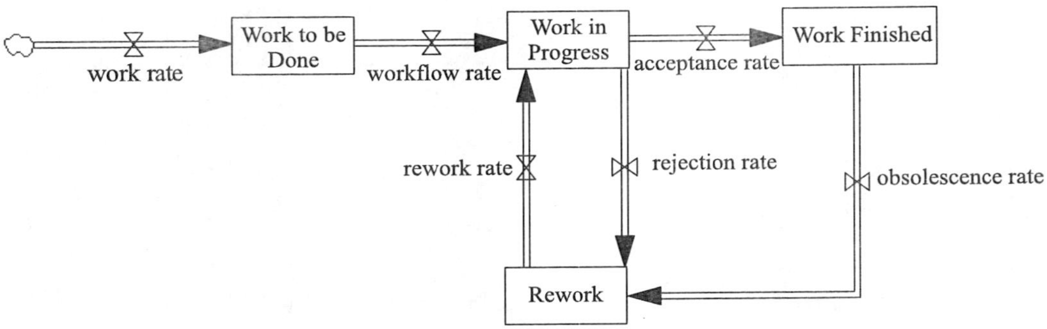

Figure 4.3: Workflow Sector (Skeleton Form).

Further consideration will show that a number of factors influence the workflow sector as shown in the detailed stock and flow diagram (Figure 4.4).

2. *Competent product development staffs sector*

The term 'competence' in the present context refers to the knowledge, skills and attitude of the PDSs that would enable them to complete the job, whereas, 'competence loss' refers to the knowledge, skill and attitude of the PDSs that go unutilized in the organization.

The stock – Potential project staffs pool, is based on the controlling factor hiring rate. These project staffs need to be trained to be the competent project staff. Hence, the training rate controls the stock - Competent project staffs pool. Further, it has to be noted that the Competent project staffs pool if affected by quitting rate and loss in competence. Accordingly, Incompetent project staffs pool will be stocked and the controlling rate gain in competence adds back to the competent project staffs pool. This constitutes the competent product development staffs' sector in the skeleton form shown in Figure 4.5.

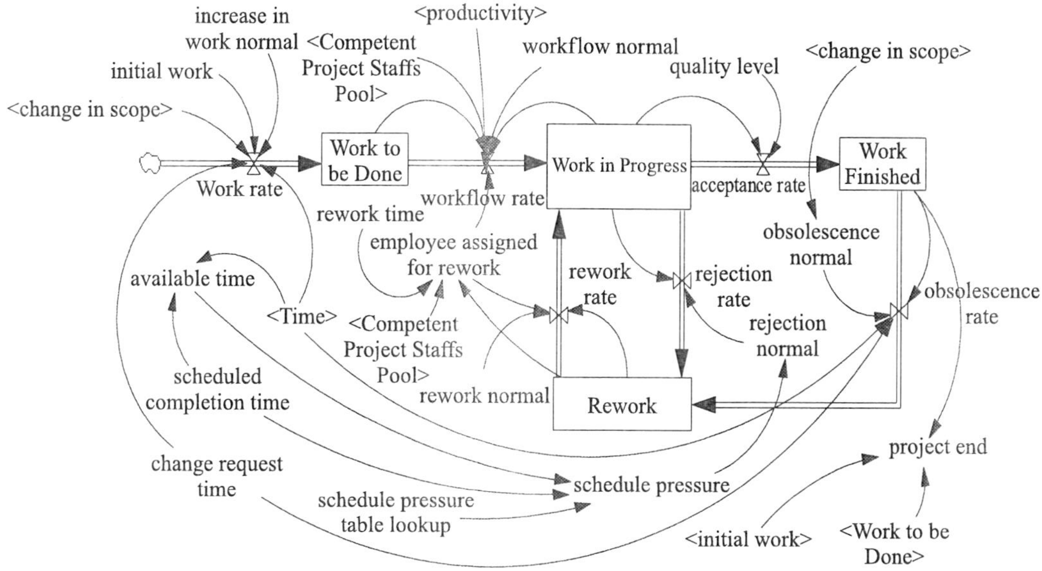

Figure 4.4: Detailed Workflow Sector.

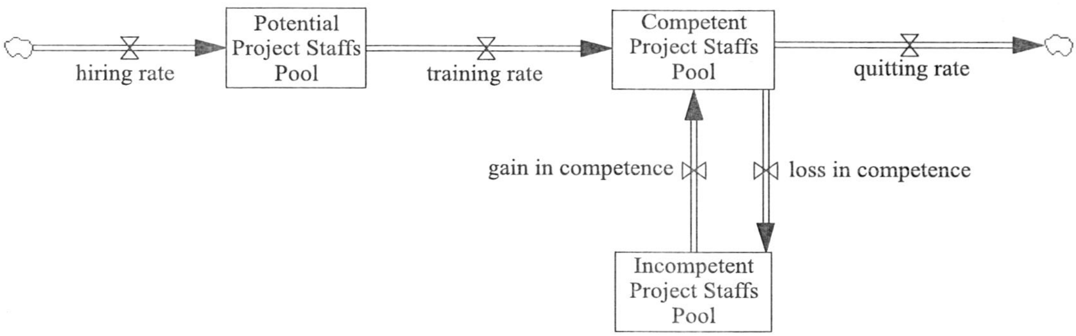

Figure 4.5: Competent Product Development Staffs' Sector (Skeleton Form).

Further, a host of variables controls the rate and stock diagram and the complete Stock and Flow diagram is shown in Figure 4.6.

Note: The variables enclosed under '<' and '>' brackets are the shadow variables, which have further set of causal relation with other variables. As the diagram becomes more and more complex shadow variables are used to make the model comprehensive as well as easily understandable. You may refer to the User's Manual of the commercially available software packages for more details on it.

3. *Product development salary and revenue sector*

Figure 4.7 illustrates product development staffs sector with salary and revenue on the focus. The work finished has a direct influence on the controlling variable revenue generated which

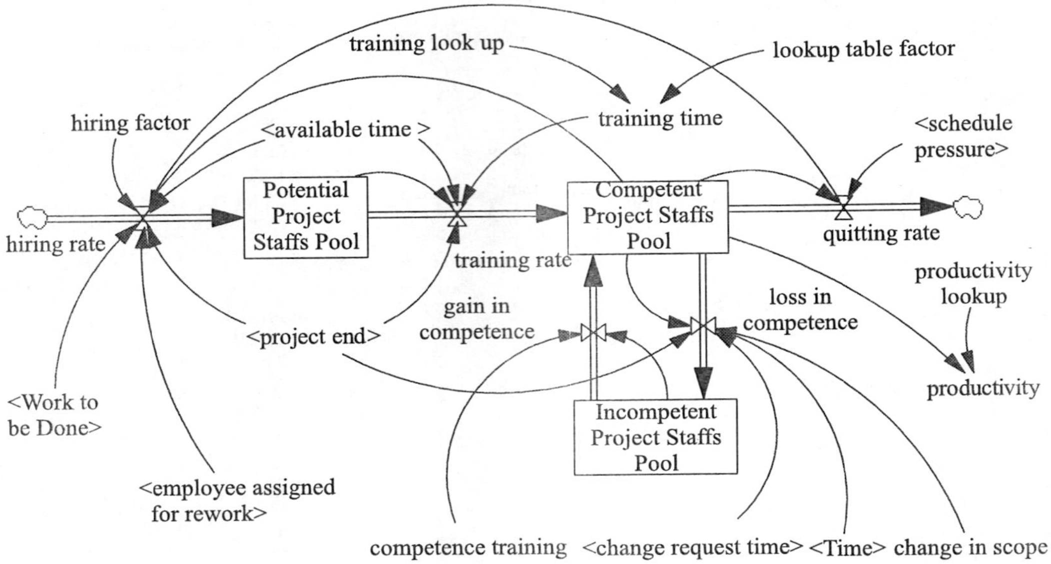

Figure 4.6: Competent Product Development Staffs' Sector.

stocks the revenue. The stock compensation is mainly in the form of salary, which is in turn influenced by several other factors as shown.

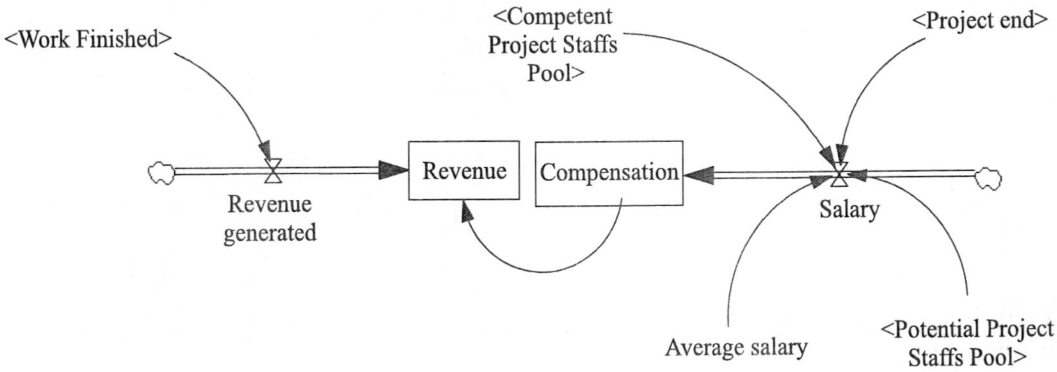

Figure 4.7: Product Development Staffs' Salary and Revenue Sector.

4.7 DYNAMIC SIMULATION

Simulation Parameters

Simulation parameters support the dynamic hypothesis by giving them the real values, which are necessary for conducting simulation. The following list gives the key simulation parameters used in this model. It has to be noted that simulation is performed basically to study the influence of 'change in scope' of the project, and hence, the values are fixed as 20%, 40%, 60% and 80% change in scope. The equations used for relating various parameters are listed in Appendix- I.

- Work to be done: 100 man-months (person*month)
- Scheduled completion time: 32 months, $D_t = 1$.
- Change is requested on the 20th month.
- Initially PDSs available: 20 (persons)
- Initial competent PDSs assigned will be 5 (persons).
- Base case-simulation is for project with no change in scope.
- Cases 1, 2, 3, 4 are for projects with 20%, 40%, 60% and 80% change in scope.
- Interaction effect is provided by way of table look up (Fried, 1995)
- Each change entails in 20% increase in workload.
- Competent PDSs requirements are adjusted on monthly basis.
- PDSs and incompetent PDSs need training time value 1 month and 0.5 month respectively.
- Competent PDSs are taken on the basis of productivity table look up
- Rework identified gets priority over project work to be done.
- All rework identified is assumed to be reworked within one month.
- Competent PDSs are allocated to rework first and the remaining are assigned to project work to be done.
- Reduction in available time increases schedule pressure ranging from 1 to 2 and this schedule pressure affects the rejection rate as well as competent PDSs turnover.
- Average wage per PDS per month is \$ 500; initial capital value invested \$ 2,000,000.
- Available PDSs at all times needs to be maintained at 10% of the project staff.
- If-then-else logic is employed in many equations to prevent negative draining of stocks as well as to prevent recurring fractional computations.

Note: The amounts indicated are in dollars, but applicable to the Indian context, e.g., salary.

4.8 RESULTS AND DISCUSSION

1. Product Development Workflow Sector

The simulation is performed to study workflow behaviour under the condition of 'change in scope'. Hence, the workflow behaviour under no change in scope is to be compared with change in scope.

Figure 4.8 shows the simulation result with change in scope varied from 20-80% in steps of 20%.

With no change in scope (0% variation in scope), the initial linear variation turns to exponential form, and it can be observed that after 24 months the work finished reaches stabilization at 100 units/person/month. But when scope changes by 20% from normal, a sudden dip is observed and the system takes about 5 months to stabilize and after 26th month, the work finished reaches a stable value of about 105 units. Similar trend is observed for the increase in the change in scope, and higher the change in scope, longer will be the time period to stabilize, but larger will be the number of work finished. For 80% change in scope, it can be observed that the system takes 12

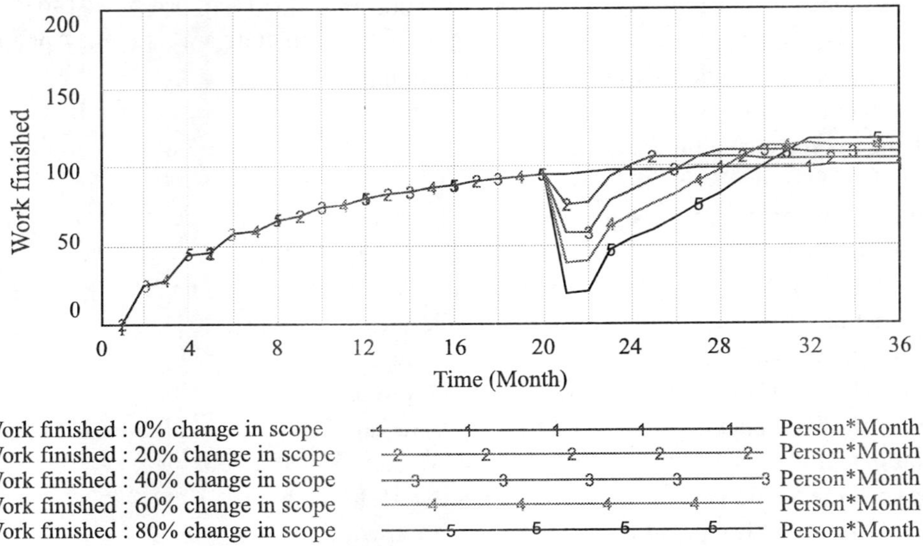

Figure 4.8: Work Finished Behaviour.

months to achieve stability, and reaches a constant production of about 120 units of finished work thereafter.

2. *Competent Product Development Staffs Sector*

Under ideal conditions, when there is no change in scope, the competent product development staffs number increases linearly and stabilizes at 10 in the 32nd month. However, under actual situation, when there is change in scope, there is a linear dip (20th month) and rise (21st month) and thereafter increases linearly up to a certain point, after which, once again starts declining as shown in Figure 4.9. This is because higher the change in scope greater will be the loss in competence, and accordingly, higher will be the competence training and hence, the gain in competence. It is interesting to note that higher the change in scope, greater will be the dip in the competent project staffs number, but also, higher will be maximum number of competent project staffs.

3. *Product Development Salary and Revenue Sector*

It can be observed that when there is no change in scope, the revenue remains constant in the first couple of months initially, and then increases gradually in a nonlinear form (Figure 4.10). But after reaching a maximum value of about 4.8 million, starts declining gradually. But due to change in scope, it starts declining after the 21st month, however, catches up gradually and starts increasing. It can be observed that the higher the change in scope, the larger will be the revenue drop.

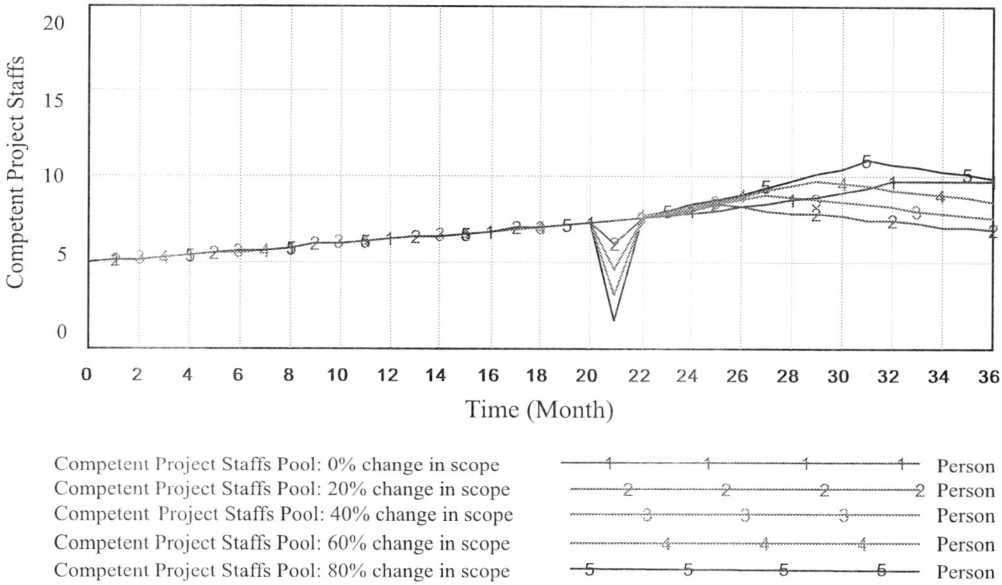

Figure 4.9: Competent PDSs Pool Behaviour.

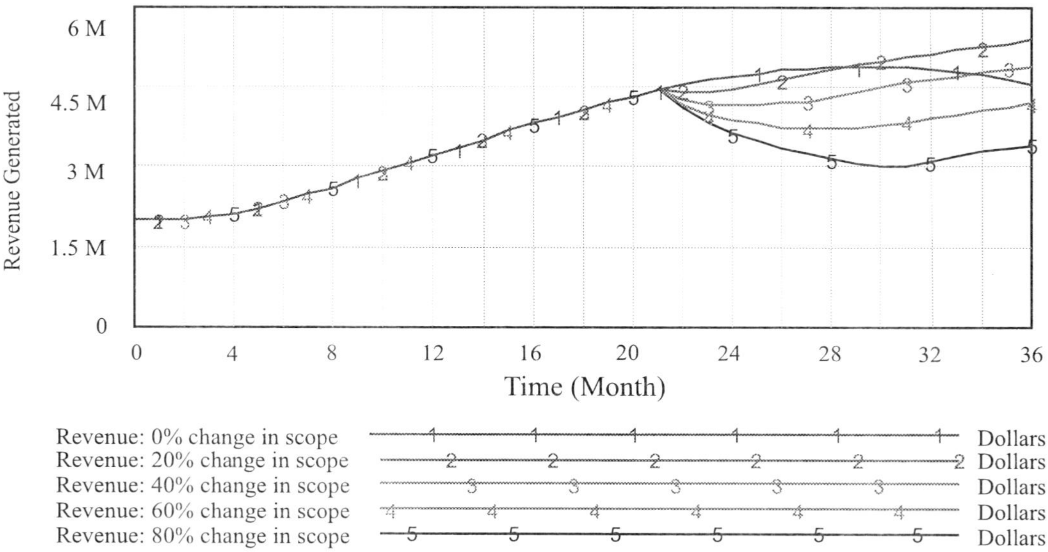

Figure 4.10: Revenue Behaviour.

4.9 INFERENCE AND IMPLICATIONS

The findings through the NPD dynamic behaviour is that the higher the 'change in scope':

- more will be the work addition and longer would be the duration to catch up with the optimum production,
- the higher the competence level generation, and
- the lesser will be the revenue generated.

Nevertheless, from the holistic point of view, even if the revenue generated is low for a given amount of work, in a particular project, the overall competence pool of the company is increased to a considerable amount, which becomes a strategic asset to the company. It is very important to note that the revenue generation has a declining trend with no change in scope, whereas an increasing trend even under maximum change in scope (Figure 4.10). Hence, beyond 36 months, the revenue generation will catch up with the project with no change in scope (curve 1) and after the point of intersection, it will continuously increase. This is exactly where the experience of a manager and his ability to plan for sustainable development of the company will come to use. Therefore, a system dynamics based simulation results should only substantiate managerial decisions and they cannot be the only criterion for decision-making.

Based on this study, the Project Manager suggests the CEO of the company that on long-term perspectives of the company in focus, 'competence development' of the PDSs should be the priority of the company, as the simulation results indicate that on a longer run it enhances the optimum work finish level and also improves the revenue.

4.10 SUMMARY

In this chapter, a case study on the Project Management in New Product Development has been taken to explain how System Dynamics can be used as a tool, and its systematic methodology can be applied to seek solution to a given problem. Starting from the basis building blocks, the systematic approach to produce simulation, and then, the interpretation of the graphs obtained has been explained so that the reader may acquaint himself/herself with the procedure. The reader should understand that in a system dynamics model, the 'numbers' are not very important as the researcher is basically interested in studying the 'trend and behaviour' of the dynamic system. However, the model must also pass the 'reality check', so the numbers used should not be unrealistic. A separate chapter given later on 'validation of a system dynamics model' lays emphasis on different ways in which a model may be validated before accepting its results, so it is premature to discuss that issue at this stage.

4.11 EXERCISE

For the model discussed in this chapter, conduct simulation to future months and find out the time period, after which, the revenue with maximum change in scope (80% change) crosses that with no change in scope (0% change). What will be the managerial implications?

5

National Engineering Competence Pool Development and Deployment

Objectives

After going through this chapter you will be able to:

- Understand the importance of National Engineering Competence Pool Planning.
- Identify the parameters which influence the Engineering Competence Pool Development.
- Construct the causal loop diagram and stock and flow diagram for the above situation.
- Appreciate the purpose and use of 3-D surface simulations.
- Interpret the simulation results and draw implications in the context of the case study under consideration.

Abbreviations and Notations:

AECPL	=	Actual Engineers' Competence Pool Level.
ALCA	=	Actual Level of Competence Absorbed.
CPG	=	Competence Gap.
CTCRATE	=	Competence Training Completion Rate.
CTRATE	=	Competence Training Rate.
DECPL	=	Desired Engineers' Competence Pool Level.
FGCFR	=	Forecast Graduate Competence Fail Rate.
FSKFR	=	Forecast Skill Fail Rate
GECOPM	=	Graduate Engineers' Competence Pool Model.
GET	=	Graduate Engineering Training
HRM	=	Human Resource Management.
KM	=	Knowledge Management.
PECOPM	=	Practicing Engineers' Competence Pool Model.

PGCFR	=	Present Graduate Competence Fail Rate.
PSKFR	=	Present Skill Fail Rate
RCRATE	=	Recruitment Completion Rate
RRATE	=	Recruitment Rate
SKG	=	Skill Gap
T_a	=	Average time to determine the forecast competence loss rate.
TCRATE	=	Training Completion Rate
T_{dt}	=	Competence Training Delay Time.
T_r	=	Time over which the competence gap is to be recovered.
TRATE	=	Training Rate
T_{rd}	=	T_{dr} = Recruitment Delay Time.
T_s	=	Skill Gap Recovery Rate
T_{sl}	=	Skill Loss Rate Averaging Time.
T_{td}	=	Training Delay Time

5.1 THE PROBLEM STATEMENT

Engineering Competence Development is in the forefront of the national agenda owing to the fact that it has a direct bearing on the economy of the country. Engineers from various disciplines from over 1000 colleges in the country are absorbed not only in the design and manufacturing sectors, but also in the software sector that is booming ever since its inception. The ideal situation would be to absorb the entire competence pool in totality, i.e., every student who joins engineering successfully graduates and every graduate passes the placement interviews and successfully completes the training in the workplace. But this is not practicable as students fail during graduation, placement interviews and also during the training period. However, the failed students will reappear for the exams, graduates keep taking placement tests until they secure a job, and some may have to join retraining if they fail in the first attempt. For example, in the software sector after absorbing the graduates they will be trained for a period of 3 months and they will be given only one chance to reappear for the test, if they don't qualify in the first attempt.

So, the above referred competence loss results in 'time lags', which are detrimental to the growth of the country. The crux of the problem lies in the recovery of 'competence loss' through retraining. So, the National Planning Commission has outsourced a project to study the critical time-based policy parameters, which influence the engineering competence pool development and deployment and identify the most critical one so as to enable the Policy Design and Implementation.

5.2 OBJECTIVES

In order to solve the above problem, following objectives are to be accomplished:

- Conceptualize the problem into a System Dynamics model.
- Identify the time-based policy variables which influence the dynamics of the system.
- Develop the causal loop and stock and flow diagrams so as to build the dynamic hypothesis.
- Conduct simulation to identify the critical time based policy parameters which need to be controlled in order to minimize the 'competence gap'.

5.3 METHODOLOGY

The methodology to be adopted may be problem specific; nevertheless, the general steps remain the same. The diagram shown in Figure 5.1 shows the methodology adopted to solve the given problem. It may be noted that the steps may be sequential in nature but some of them may be interlinked but all the steps are oriented towards the 'understanding of the system'. This is because, at each stage of the System Dynamics procedure one may gain a better insight into the problem and may lead to better understanding of the problem.

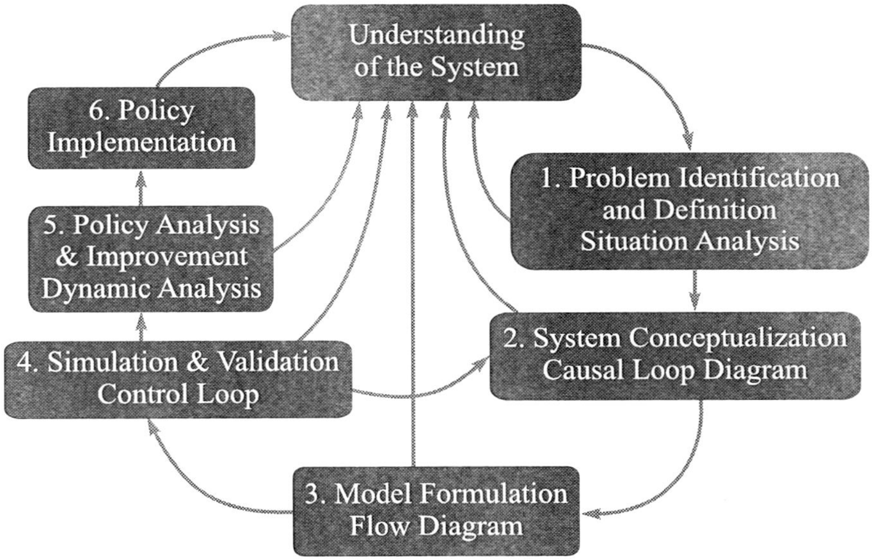

Figure 5.1: The Research Methodology.

5.3.1 Problem Identification

To start with, the situation analysis has to be performed, which includes problem identification, i.e., the study of the number of engineers produced in the country per year and their absorption rate. The variables relating to the problem situation have to be filtered out and related to each other to find out the causal relationships and feedbacks in the system. The crucial part of problem identification is to identify the time-based policy parameters, which have influence on the system dynamics.

The following are the six time-based policy parameters of this model;

T_r = Time over which the competence gap is to be recovered,
T_{dt} = Competence training delay time,
T_a = Average time to determine the forecast competence loss rate,
T_s = Skill gap recovery rate,
T_{td} = Industrial training delay time, and
T_{sl} = Skill loss rate averaging time.

The term 'competence' refers to the knowledge, skill and attitudes of the engineers/future engineers that would enable them to be employed in the global market, join for higher studies, or become entrepreneurs. Competence loss refers to the failure of the students/engineers to be qualifying as graduates, being employable, pursue higher studies, or become entrepreneurs.

5.3.2 System Conceptualization

The parameters that have significant influence on competency development of the engineers have to be identified in this stage.

The fundamental equation:

$$Gap = Desired - Actual$$

This forms the starting point of the causal diagram (Figure 5.2). 'Competence Gap' is the key parameter which is causally related to 'the desired engineer's competence pool level' and 'the actual engineers' competence pool level'. The other parameters are related as shown in Figure 5.2. The model is called the Engineering Competence Pool Model (ECOPM) as it basically studies the System Dynamics of Engineering Competence Pool.

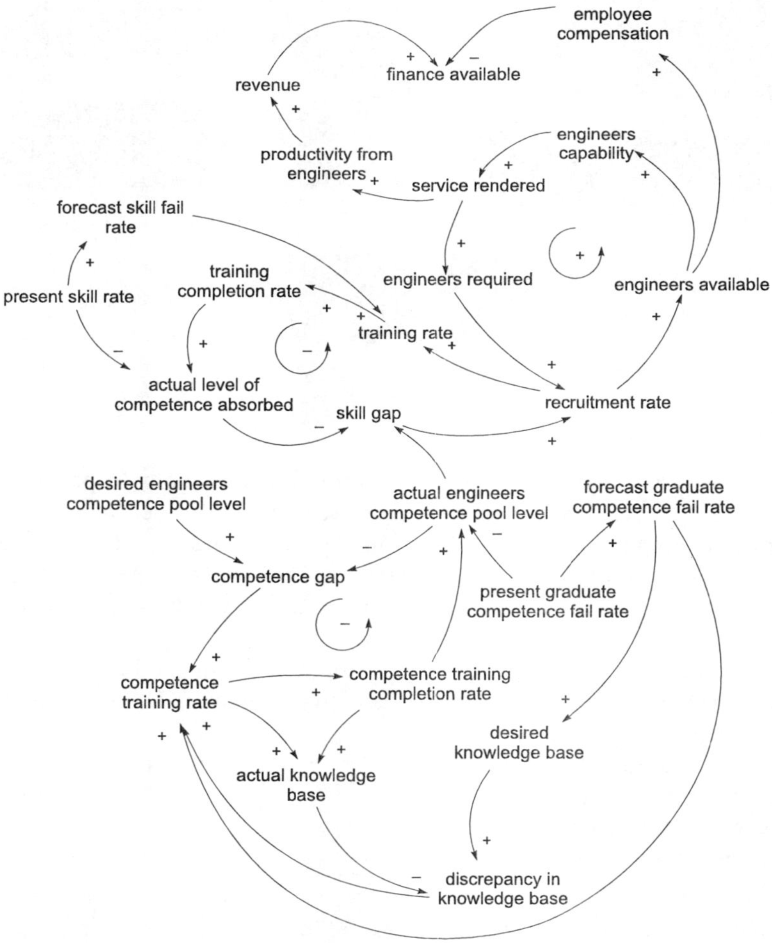

Figure 5.2: Engineering Competence Pool Model Causal Diagram.

5.3.3 Model Formulation

The next stage is to formulate the model and develop the Flow Diagram. Figure 5.3 illustrates the feedback control (based on competence gap) and feed-forward control (based on forecast competence loss rate). In order to analyze the dynamic response, the time-based policy variables have been considered.

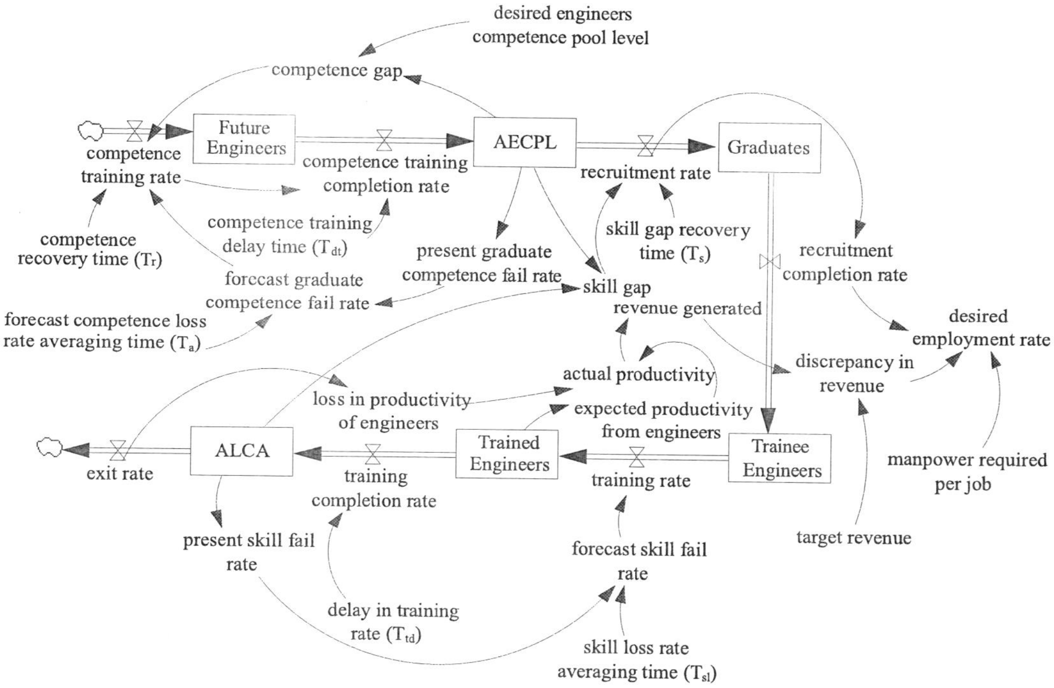

Figure 5.3: Engineering Competence Pool Model—Stock & Flow Diagram.

The governing equations giving discrete-time feed-forward and feedback difference equations used to calculate the system parameters are given below:

$$CPG = DECPL - AECPL \quad \ldots[1]$$

$$CTRATE = CPG\left(\frac{1}{T_r}\right) - FGCFR \quad \ldots[2]$$

$$CTCRATE = CTRATE*\left(\frac{1}{1+T_{dt}*S}\right) \quad \ldots[3]$$

$$AECPL = \frac{1}{S}*(CTCRATE - PGCFR) \quad \ldots[4]$$

$$FGCFR = PGCFR*\left(\frac{1}{1+T_a*S}\right) \quad \ldots[5]$$

$$SKG = AECPL - ALCA \quad \ldots[6]$$

$$RRATE = SKG^* \frac{1}{T_s} \quad ..[7]$$

$$RCRATE = RATE^* \frac{1}{1 \quad T_{rd} * S} \quad ..[8]$$

$$TRATE = RCRATE + FSKFR \quad ..[9]$$

$$TCRATE = TRATE^* \frac{1}{1 \quad T_{td} * S} \quad ..[10]$$

$$ALCA = \frac{1}{S} * (TCRATE - PSKFR) \quad ..[11]$$

$$FSKFR = PSKFR^* \frac{1}{1 \quad T_{sl} * S} \quad ..[12]$$

At a first glance the stock and flow diagram looks a bit complicated but on analysis it follows a simple logic. The engineering students are referred to as future engineers who successfully pass the competence training. Here competence training refers to the imparting of the knowledge, skill and attitude required for a student to qualify himself/herself into a successful engineering graduate. Hence, the competence training completion rate will decide the stock actual engineering competence pool level (AECPL). It is this AECPL that is subjected to recruitment, and hence, the recruitment rate controls the total stock of the successful graduates. Again, the recruitment completion rate will govern the stock of the number of trainee engineers. These trainee engineers are usually subjected to some training in the workplace, e.g., in IT sector the general period is 3 months, and the training rate controls the stock trained engineers. Finally, the training completion rate will control the stock actual level of competence absorbed (ALCA). The exit rate of engineers (attrition) will drain the ALCA.

The main aim of this study is to identify the time-based policy parameters which have influence on the Engineering Competence Pool. Hence, the most crucial part of this study is to identify the critical time-based policy parameters. Six such time-based policy parameters can be identified which significantly influence the System Dynamics of Engineering Competence Pool Model:

1. Competence Recovery Time (T_r)

Those who fail to acquire the required amount of engineering competence will add to the national competence loss. Hence, additional time is required to recover the loss in this competence gap. The time taken to recover this lost competence is referred to as Competence Recovery Time.

2. Competence Training Delay Time (T_{dt})

During the competence training for those who fail to acquire the necessary skill, again there could be further delay during the training period itself which may demand retraining. This is termed Competence Training Delay Time.

3. Forecast Competence Loss Rate Averaging Time (T_a)

On a timely basis, the competence fail rate of graduates may be forecast and a feed-forward may be introduced to train extra number of graduates to fill the competence gap. So this forms the feed-forward loop. So the time taken to calculate the competence loss is termed Forecast Competence Loss Rate Averaging Time.

4. *Skill Gap Recovery Time (T_s)*

During the recruitment, the graduates who have passed the engineering exams may fail due to various reasons such as lack of interview facing skills, lack of aptitude, lack of subject knowledge retention ability, presentation skills, etc. So such engineers will have to reappear for the interviews or attend interviews in different companies. This again causes a delay and is termed Skill Gap Recovery Time.

5. *Delay in Training Rate (T_{td})*

The trainee engineers who fail to acquire the required amount of skills during the stipulated period (say 3 months in IT sectors) will be given an opportunity to retrain themselves (in most of the cases only one chance will be given failing which the candidate will have to leave the company) and there could be a delay here again. This is termed *Delay in Training Rate.*

6. *Skill Loss Rate Averaging Time (T_{sl})*

Based on the past statistics the skill fail rate of engineers during their training in the industries can be forecast and additional number of graduates may be recruited. The delay in this forecast is termed Skill Loss Rate Averaging Time.

So, in the stock and flow diagram of the ECOPM model, the six time-based policy parameters have been identified, whose influence on the Engineering Competence Pool Development and Deployment has to be studied.

MATLAB® Simulink Software has been used to develop the control loop of the stock and flow diagram and is given in Figure 5.4. Even though other software can be used to serve the purpose as 3D-surface graphs are required in this case MATLAB has been used in this case. A preliminary knowledge on Control Theory and Laplace Functions is desirable and any book which teaches these fundamentals would be sufficient to develop the control loop and conduct simulation.

5.3.4 Initial Conditions

In any System Dynamics analysis, we need the initial values to start the simulation. It is to be noted that in System Dynamics Simulation it is the trend analysis which is given importance and numbers do not have significance, however, the numbers should be, as far as possible, close to real life situations. In the present context, following initial values have been assumed based on the real life situations:

- Desired Engineering Competence Pool Level (DECPL) = 250,000 units.
- The delay time in training (T_{dt}) is taken to be 6 months (one semester) to enable them to get the required competence level through re-training.
- The drop out is about 5% of engineers.
- The delay in training rate is 3 months.
- The skill loss is assumed to be 5%.

It can be observed that in the first part of the control loop the Desired engineering competence pool level (DECPL) is matched with the Actual engineering competence pool level (AECPL).

Hence, a feedback to correct the discrepancy is brought into the summing joint indicating the competence gap (CPG). The competence training rate (CTRATE) is calculated by dividing the discrepancy between the desired and actual value of the level by a time factor (T_r), which represents the average delay in performing the training rate. The competence training completion rate (CTCRATE) is influenced by the *competence training delay time* (T_{dt}) as shown in the Laplace function. Further the feedback and feed-forward influenced by the forecast competence loss rate averaging time (T_a) gives the forecast graduate competence fail rate (FGCFR) and the present graduate competence fail rate (PGCFR). The output from this loop gives the summation as actual engineering competence pool level (AECPL). Similar loop continues to the transformation of actual engineering competence pool level (AECPL) to the practising engineers based on the actual level of competence absorbed (ALCA) and the exit rate which follows thereafter.

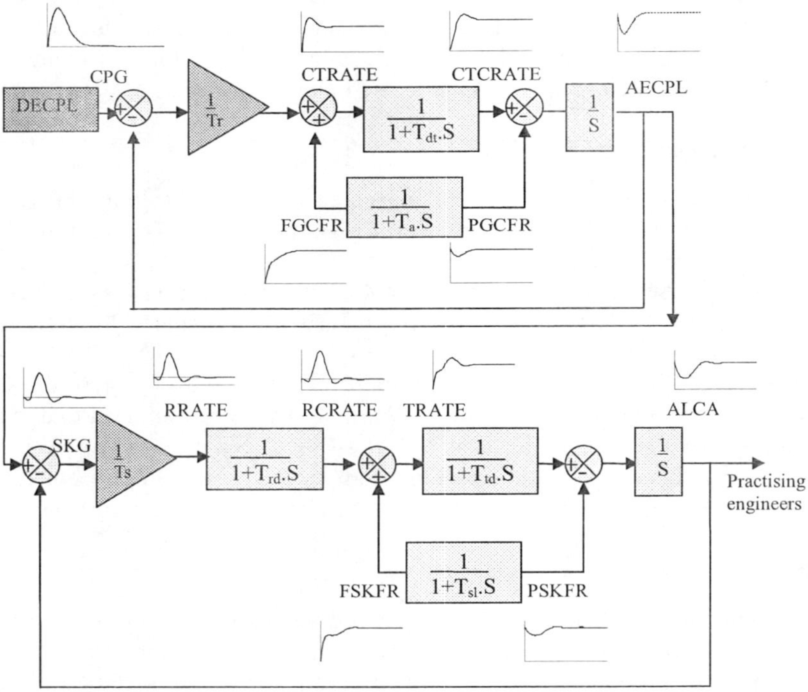

Figure 5.4: Control Loop of ECOPM Model.

5.3.5 Computer Simulation

During the ideal operation, the number of graduates produced equals the number of graduates absorbed in the market. But the problem arises when all the admitted graduates fail to acquire the required level of competence. Also, the market requirement for skilled engineering graduate changes due to several governing factors, as a result of which, the recruitment may be denied/ delayed. So the time-based policy parameters, viz., T_r, T_{dt}, T_a, T_s, T_{td} and T_{sl} are to be varied, one at a time, to determine their optimum setting. The influence of these parameters on the Recruitment Completion Rate, Training Completion Rate, and Actual Level of Competence Absorbed would provide useful input for policy-makers.

With the initial conditions given in the previous stage, simulations may be performed to study the system behaviour. Following are the influences of the six time-based policy parameters.

1. Competence Gap Recovery Time (T_r)

(a) *Influence on Recruitment Completion Rate (RCRATE)*

Simulation Parameters:

- $T_{dt} = T_a = 6$ months (Maximum period for competency recovery).
- T_r (Competency recovery rate) is varied from 0 to 30 months.
- $T_{rd} = T_{td} = T_{sl} = T_s = 3$ months

Simulation result is shown in Figure 5.5.

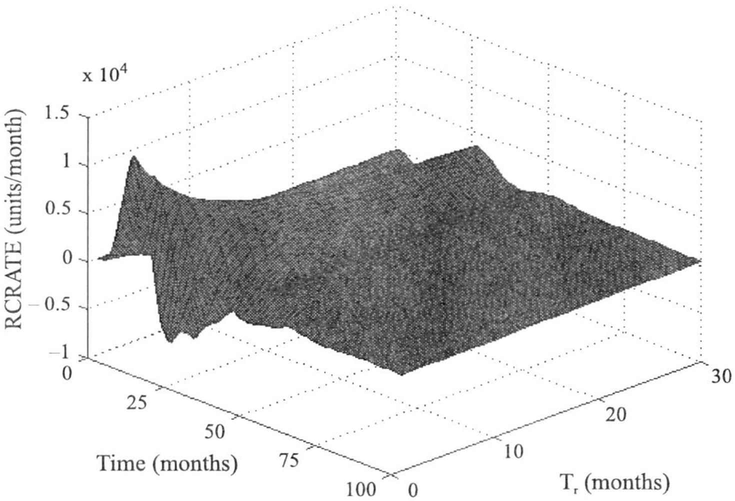

Figure 5.5: Response of RCRATE for T_r.

Inference:

- Higher T_r values increase the time for rise of the Recruitment Completion Rate. Higher T_r values also indicate that the rate of people being recruited has decreased. The settling time for higher values of T_r is low.
- The system takes 60 months to settle, i.e., the system is able to recover from the fluctuations only after duration of 60 months.
- The smaller the T_r values, the lesser the time taken for the RCRATE to rise. Smaller T_r values also indicate that the rate of people being recruited is more. Settling time for lower values of T_r is high.

For better results, optimum value of T_r should be ranging from a minimum of 3 to a maximum of 6 months.

(b) Influence on Training Completion Rate (TCRATE)

Simulation Parameters:

- $T_{dt} = T_a = 6$ months (Maximum period for competency recovery).
- T_r (Competency recovery rate) is varied from 0 to 30 months.
- $T_{rd} = T_{td} = T_{sl} = T_s = 3$ months

Simulation result is shown in Figure 5.6.

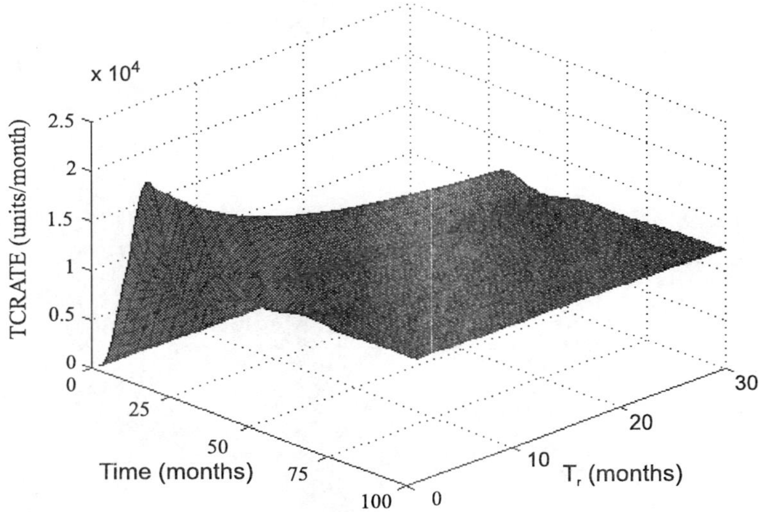

Figure 5.6: Responce of TCRATE for T_r.

Inference:

- Higher T_r values increase the time for rise of the Training Completion Rate. Higher T_r values also indicate that the rate of people being trained is low. Settling time for higher values of T_r is low.
- The system takes 65 months to settle, i.e., the system is able to recover from the fluctuations only after a duration of 65 months.
- The smaller the T_r values, the lesser the time taken for the Training Completion Rate to rise. Smaller T_r values also indicate that the rate of people being trained is high. The settling time for lower values of T_r is high.

For better results, the optimum value of T_r should be ranging from a minimum of 3 months to maximum of 6 months.

(c) Influence on Actual Level of Competence Absorbed (ALCA)

Simulation Parameters:

- $T_{dt} = T_a = 6$ months (Maximum period for competency recovery).
- T_r (Competency recovery rate) is varied from 0 to 30 months.
- $T_{rd} = T_{td} = T_{sl} = T_s = 3$ months

Simulation result is shown in Figure 5.7.

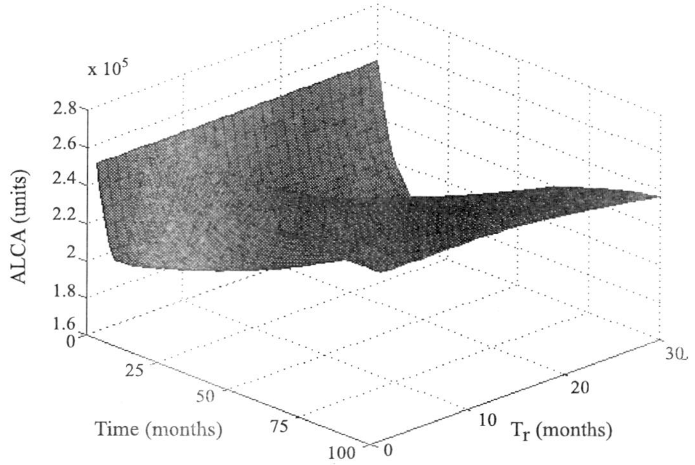

Figure 5.7: Response of ALCA for T_r.

Inference:

- The larger the T_r, values, the larger the competence pool drop, indicating that the system is unable to recover from the competency shortages over a period of time.
- The competence pool takes 40 months to settle, i.e., the system is able to recover from the competency shortage only after a duration of about 40 months for higher values of T_r.
- Lower values of T_r indicate a smooth rise in the competence level for values of T_r up to the first 5 months. This smooth rise is then followed by oscillations after which the system response remains constant.

For better results, optimum value of T_r should be ranging from a minimum of 3 to maximum of 6 months.

2. Competence Training Delay Time (T_{dt})

(a) Influence on Recruitment Completion Rate (RCRATE)

Simulation Parameters:

- $T_r = T_a = 6$ months (Maximum period for competency recovery).
- T_{dt} (Competence Training Delay Time) is varied from 0 to 30 months.
- $T_{rd} = T_{td} = T_{sl} = T_s = 3$ months

Simulation result is shown in Figure 5.8.

Inference:

- Higher T_{dt} values increase the time for the rise of the RCRATE. Higher T_{dt} values also indicate that the rate of people being recruited has increased. The settling time for higher values of T_{dt} is high.
- The system takes 60 months to settle, i.e., the system is able to recover from the fluctuations only after a duration of 60 months.

- The smaller the T_{dt} values, the lower the time taken for the RCRATE to rise. Smaller T_{dt} values also indicate that the rate of people being recruited has decreased. The settling time for lower values of T_{dt} is low.

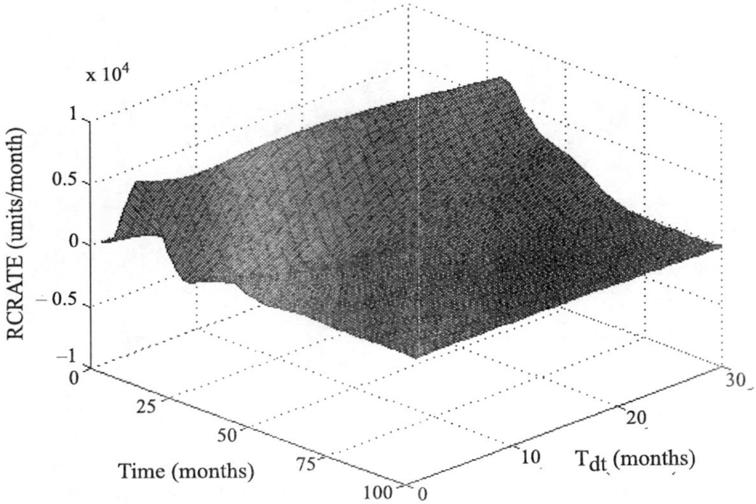

Figure 5.8: Response of RCRATE for T_{dt}.

For better results, optimum value of T_{dt} should be ranging from a minimum of 3 to a maximum of 5 months.

(b) Influence on Training Completion Rate (TCRATE)

Simulation Parameters:

- $T_r = T_a = 6$ months (Maximum period for competency recovery).
- T_{dt} (Competence Training Delay Time) is varied from 0 to 30 months.
- $T_{rd} = T_{td} = T_{sl} = T_s = 3$ months

Simulation result is shown in Figure 5.9.

Inference:

- Higher T_{dt} values increase the time for rise of the Training Completion Rate. Higher T_{dt} values also indicate that the rate of people being trained is low. The settling time for higher values of T_{dt} is high.
- The system takes 65 months to settle, i.e., the system is able to recover from the fluctuations only after a duration of 65 months.
- The smaller the T_{dt} values, the lesser the time taken for the Training Completion Rate to rise. Smaller T_{dt} values also indicate that the rate of people being trained is high. The settling time for lower values of T_{dt} is low.

For better results, optimum value of T_{dt} should be ranging from a minimum of 5 to a maximum of 8 months.

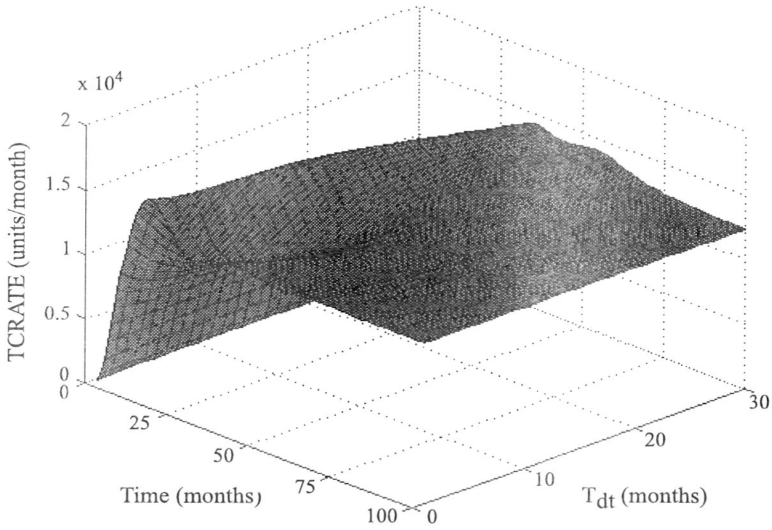

Figure 5.9: Response of TCRATE for T_{dt}.

(c) Influence on Actual Level of Competence Absorbed (ALCA)

Simulation Parameters:

- $T_r = T_a = 6$ months (Maximum period for competency recovery).
- T_{dt} (Competence Training Delay Time) is varied from 0 to 30 months.
- $T_{rd} = T_{td} = T_{sl} = T_s = 3$ months

Simulation result is shown in Figure 5.10.

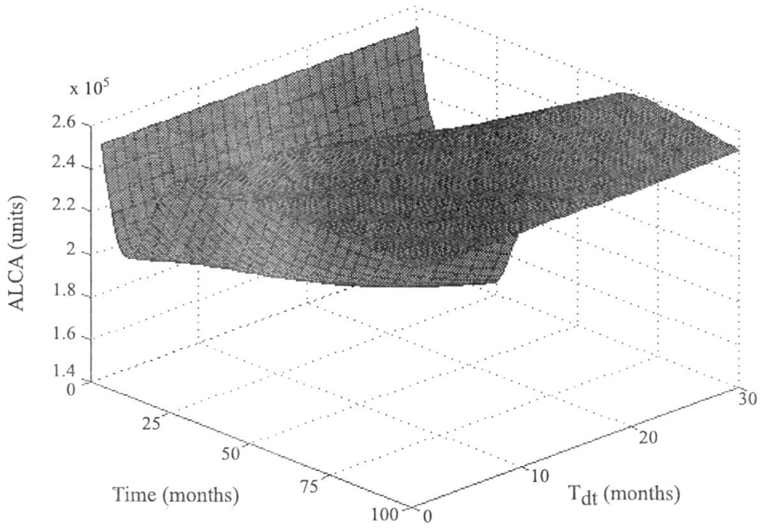

Figure 5.10: Response of ALCA for T_{dt}.

Simulation Parameters:
- $T_r = T_a = 6$ months (Maximum period for competency recovery).
- T_{dt} (Competence Training Delay Time) is varied from 0 to 30 months.
- $T_{rd} = T_{td} = T_{sl} = T_s = 3$ months

Inference:
- The larger the T_{dt} values, the larger the competence pool drop, indicating that the system is unable to recover from the competency shortage over a period of time.
- The competence pool takes 65 months to settle, i.e., the system is able to recover from the competency shortage only after a duration of about 65 months for higher values of T_{dt}.
- Lower values of T_r indicate a smooth rise in the competence level for values of T_r ranging up to the first 5 months. This smooth rise is then followed by fluctuations after which the system response remains constant.

For better results, the optimum value of T_{dt} should be ranging from a minimum of 4 to a maximum of 8 months.

3. Average time to determine the Forecast Competence Loss Rate (T_a)

(a) Influence on Recruitment Completion Rate (RCRATE)

Simulation Parameters:
- $T_{dt} = T_r = 6$ months (Maximum period for competency recovery).
- T_a (Competence Loss Rate Averaging Time) is varied from 0 to 30 months.
- $T_{rd} = T_{td} = T_{sl} = T_s = 3$ months

Simulation result is shown in Figure 5.11.

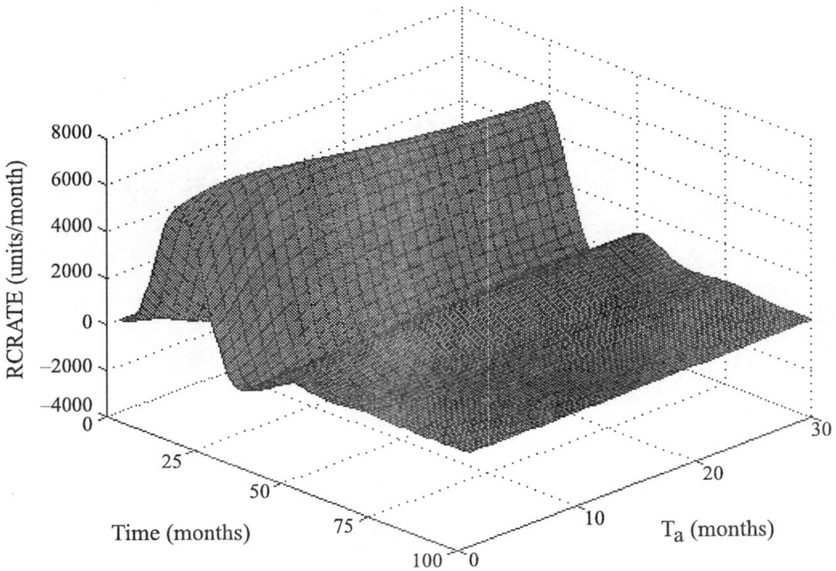

Figure 5.11: Response of RCRATE for T_a.

Inference:

- Higher T_a values slightly increase the time for rise of the RCRATE. Higher T_a values also indicate that the rate of people being recruited has decreased slightly. The settling time for higher values of T_a is slightly high.
- The system takes 60 months to settle, i.e., the system is able to recover from the fluctuations only after duration of 60 months.
- Smaller the T_a values, lesser the time taken for the RCRATE to rise. Smaller T_a values also indicate that the rate of people being recruited is more. Settling time for lower values of T_a is low.

For better results, the optimum value of T_a should be ranging from minimum of 4 to maximum of 8 months.

(b) *Influence on Training Completion Rate (TCRATE)*

Simulation Parameters:

- $T_{dt} = T_r = 6$ months (Maximum period for competency recovery).
- T_a (Competence Loss Rate Averaging Time) is varied from 0 to 30 months.
- $T_{rd} = T_{td} = T_{sl} = T_s = 3$ months

Simulation result is shown in Figure 5.12.

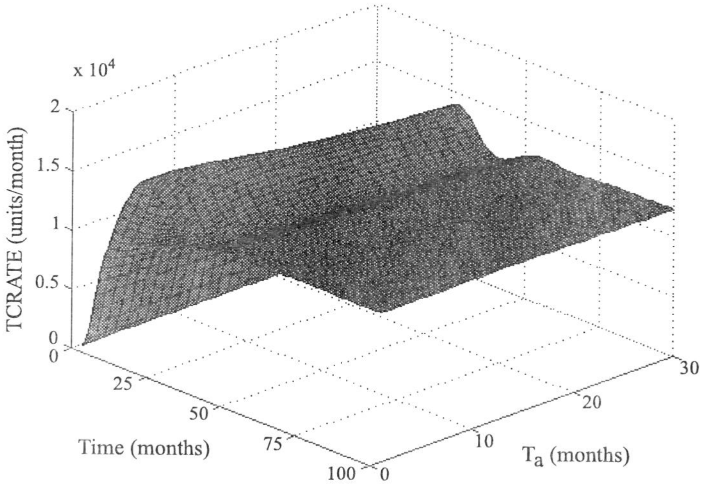

Figure 5.12: Response of TCRATE for T_a.

Inference:

- Higher T_a values slightly increase the time for the rise of the Training Completion Rate. Higher T_a values also indicate that the rate of people being trained is low. Settling time for higher values of T_a is slightly high.
- The system takes 65 months to settle, i.e., the system is able to recover from the fluctuations only after a duration of 65 months.

- The smaller the T_a values, the lesser the time taken for the Training Completion Rate to rise. Smaller T_a values also indicate that the rate of people being trained is high. The settling time for lower values of T_a is low.

For better results, the optimum value of T_a should be ranging from a minimum of one month to a maximum of 3 months.

(c) Influence on Actual Level of Competence Absorbed (ALCA)

Simulation Parameters:

- $T_{dt} = T_r = 6$ months (Maximum period for competency recovery).
- T_a (Competence Loss Rate Averaging Time) is varied from 0 to 30 months.
- $T_{rd} = T_{td} = T_{sl} = T_s = 3$ months

Simulation result is shown in Figure 5.13.

Inference:

- The larger the T_a values, the larger the competence pool drop, indicating that the system is unable to recover from the competency shortages over a period of time.
- The competence pool takes 70 months to settle, i.e., the system is able to recover from the competency shortages only after a duration of about 70 months for higher values of T_a.
- Lower values of T_a result in a smooth rise in the competence level for the first 6 months. It is then followed by a peak value after which the system response remains constant.

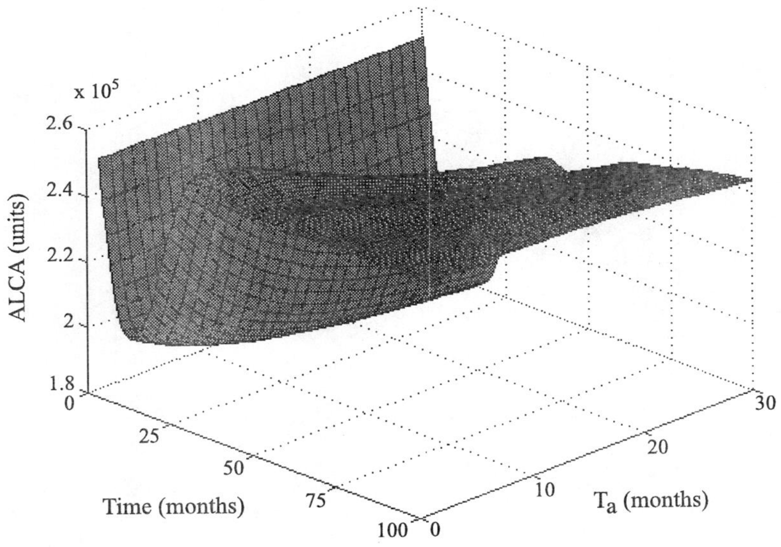

Figure 5.13: Response of ALCA for T_a

For better results, the optimum value of T_a should be about 2 months, but not more than 7 months.

4. Skill Gap Recovery Rate (T_s)

(a) *Influence on Recruitment Completion Rate (RCRATE)*

Simulation Parameters:

- $T_{dt} = T_a = T_r = 6$ months (Maximum period for competency recovery).
- T_s (Skill Gap Recovery Rate) is varied from 0 to 20 months.
- $T_{rd} = T_{td} = T_{sl} = 3$ months

Simulation result is shown in Figure 5.14.

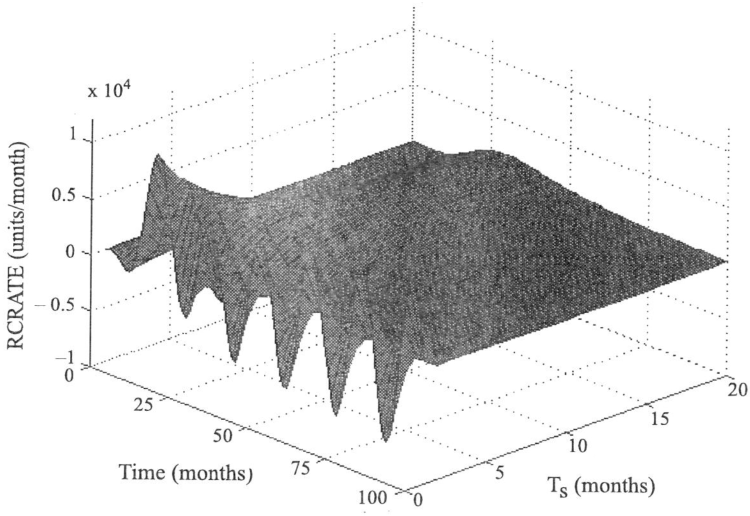

Figure 5.14: Response of RCRATE for T_s.

Inference:

- Higher T_s values increase the time for rise of the Recruitment Completion Rate. Higher T_r values also indicate that the rate of people being recruited has decreased. Settling time for higher values of T_s is low.
- The system takes 38 months to settle for values of T_s ranging from 5 to 20 months.
- The smaller the T_s value, lesser will be the time taken for the RCRATE to rise. No settling of Recruitment Completion Rate for lower values of T_s, i.e., up to first 5 months. For values of T_s ranging from 6-10 months, the system takes about 48 months to settle.

For better results, the optimum value of T_s should be ranging from a minimum of 4 to a maximum of 6 months.

(b) *Influence on Training Completion Rate (TCRATE)*

Simulation Parameters:

- $T_{dt} = T_a = T_r = 6$ months (Maximum period for competency recovery).
- T_s (Skill Gap Recovery Rate) is varied from 0 to 20 months.
- $T_{rd} = T_{td} = T_{sl} = 3$ months

Simulation result is shown in Figure 5.15.

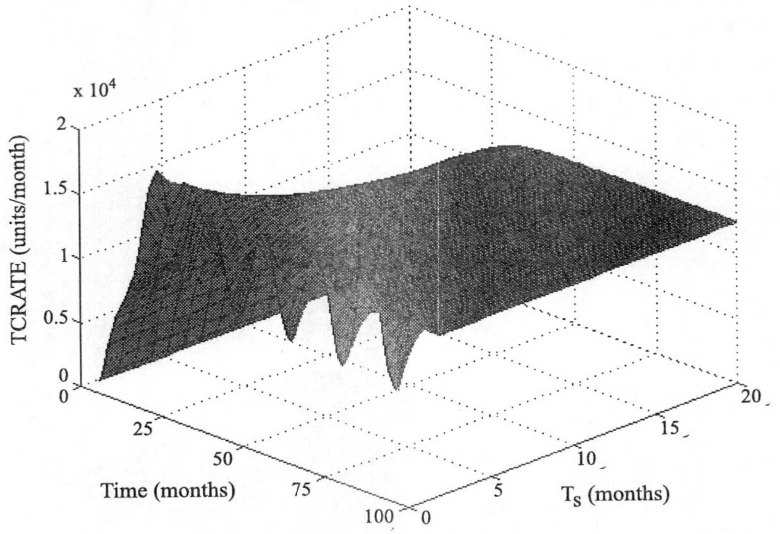

Figure 5.15: Response of TCRATE for T_s.

Inference:

- Higher T_s values increase the time for the rise of Training Completion Rate. Higher T_s values also decrease the training rate. The settling time for higher values of T_s is low.
- The system takes 30 months to settle for values of T_s ranging from 3 to 20 months.
- The smaller the T_s value, the shorter the time taken for the Training Completion Rate to rise. Smaller T_s values also indicate aggressive hiring & firing policy. No settling is observed for the Training Completion Rate for lower values of T_s, i.e., up to first 3 months. For values of T_s ranging from 4-10 months, the system takes about 30 months to settle.

For better results, optimum value of T_s should be ranging from a minimum of 4 to a maximum of 6 months.

(c) Influence on Actual Level of Competence Absorbed (ALCA)

Simulation Parameters:

- $T_{dt} = T_a = T_r = 6$ months (Maximum period for competency recovery).
- T_s (Skill Gap Recovery Rate) is varied from 0 to 20 months.
- $T_{rd} = T_{td} = T_{sl} = 3$ months

Simulation result is shown in Figure 5.16.

Inference:

- The larger the T_s values, the larger the competence pool drop, indicating that the system is unable to recover from the competency shortages over a period of time.
- The competence pool takes 78 months to settle, i.e., the system is able to recover from the competency shortages only after a duration of about 78 months for higher values of T_s.

- Lower values of T_s indicate the fluctuations in the competence level for values of T_s up to the first 3 months. For T_s values ranging beyond 3 months, there is a smooth rise in the competence pool level, followed by which the system response remains constant.

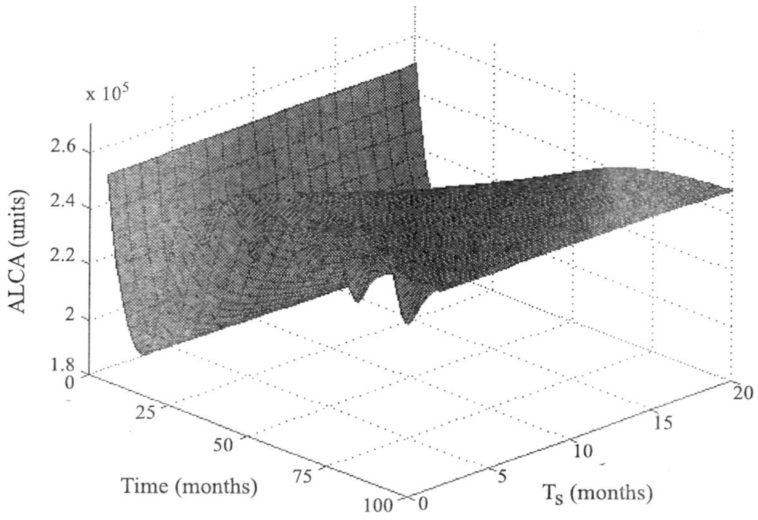

Figure 5.16: Response of ALCA for T_s.

For better results, the optimum value of T_s should be ranging from a minimum of 4 to a maximum of 6 months.

5. Industrial Training Delay Time (T_{td})

(a) Influence on Recruitment Completion Rate (RCRATE)

Simulation Parameters:

- $T_{dt} = T_a = T_r = 6$ months (Maximum period for competency recovery).
- T_{td} (Training Delay Time) is varied from 0 to 20 months.
- $T_s = T_{rd} = T_{sl} = 3$ months

Simulation result is shown in Figure 5.17.

Inference:

- Higher T_{td} values slightly increase the time for rise of the Recruitment Completion Rate. Higher T_{td} values also indicate that the rate of people being recruited is increased. The settling time for higher values of T_{td} is high.
- The system takes 80 months to settle, i.e., the system is able to recover from the fluctuations only after a duration of 80 months.
- The smaller the T_{td} values, the lesser the time taken for the RCRATE to rise. Smaller T_{td} values also indicate that the rate of people being recruited is less. Settling time for lower values of T_{td} is low.

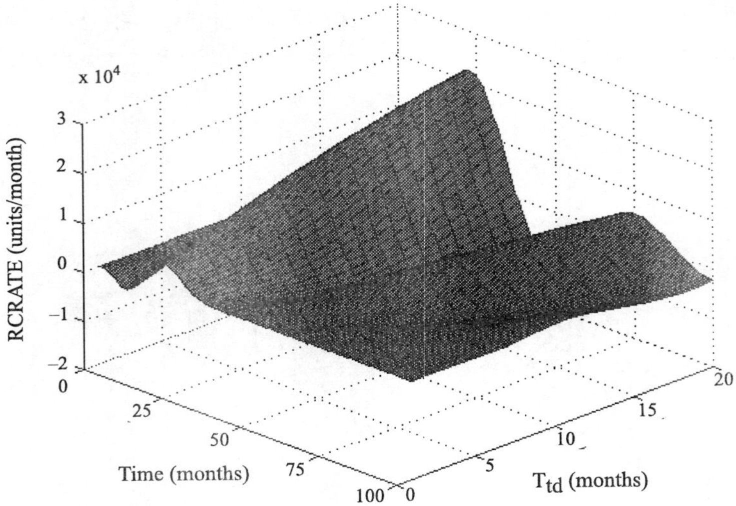

Figure 5.17: Response of RCRATE for T_{td}.

For better results, the optimum value of T_{td} should be ranging from a minimum of 3 to a maximum of 7 months.

(b) Influence on Training Completion Rate (TCRATE)

Simulation Parameters:

- $T_{dt} = T_a = T_r = 6$ months (Maximum period for competency recovery).
- T_{td} (Training Delay Time) is varied from 0 to 20 months.
- $T_s = T_{rd} = T_{sl} = 3$ months

Simulation result is shown in Figure 5.18.

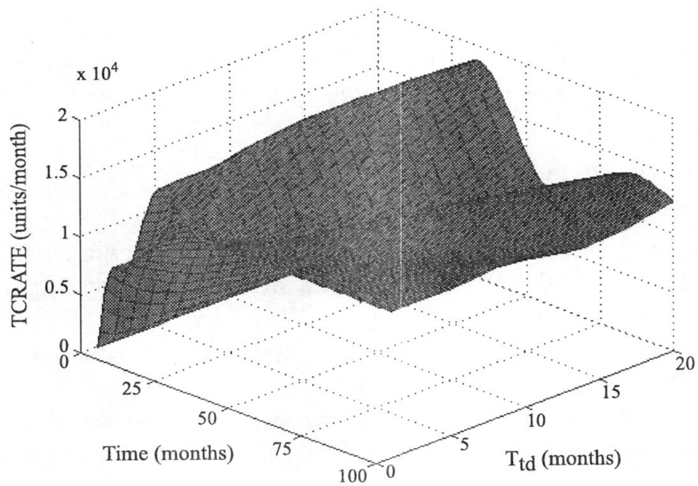

Figure 5.18: Response of TCRATE for T_{td}.

Inference:

- Higher T_{td} values slightly increase the time for the rise of the Training Completion Rate. Higher T_{td} values also indicate that the rate of people being trained is high. The settling time for higher values of T_{td} is high.
- The system takes 80 months to settle, i.e., the system is able to recover from the fluctuations only after duration of 80 months.
- The smaller the T_{td} values, the lesser the time taken for the Training Completion Rate to rise. Smaller T_{td} values also indicate that the rate of people being trained is less. The settling time for lower values of T_{td} is low.

For better results, optimum value of T_{td} should be ranging from a minimum of 2 to maximum of 7 months.

(c) Influence on Actual Level of Competence Absorbed (ALCA)

Simulation Parameters:

- $T_{dt} = T_a = T_r = 6$ months (Maximum period for competency recovery).
- T_{td} (Training Delay Time) is varied from 0 to 20 months.
- $T_s = T_{rd} = T_{sl} = 3$ months

Simulation result is shown in Figure 5.19.

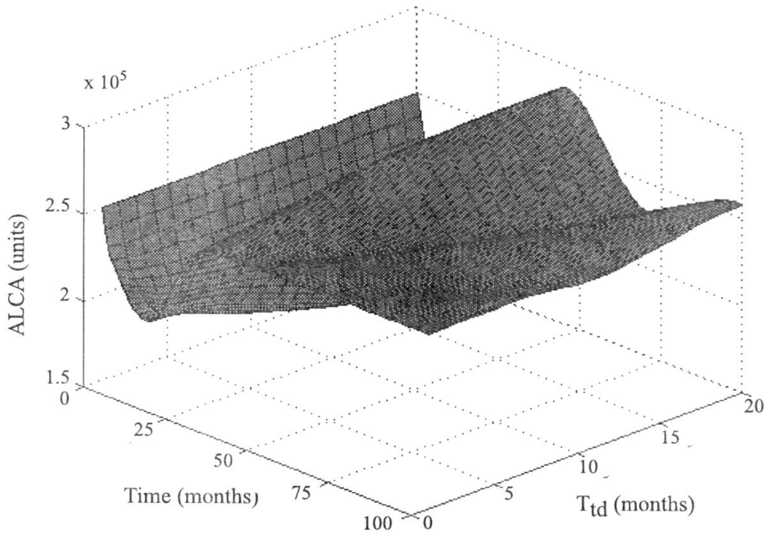

Figure 5.19: Response of ALCA for T_{td}.

Inference:

- The larger the T_{td} values, the larger the competence pool drop, indicating that the system is unable to recover from the competency shortage over a period of time.

- The competence pool takes 90 months to settle, i.e., the system is able to recover from the competency shortages only after a duration of about 90 months for higher values of T_{td} .
- Lower values of T_{td} indicate a smooth rise in the competence level for values of T_{td} up to the first 4 months. A smooth rise is the followed by peak value after which the system response remains constant.

For better results, the optimum value of T_{td} should be ranging from a minimum of 2 to a maximum of 6 months.

6. Skill Loss Rate Averaging Time (T_{sl})

(a) Influence on Recruitment Completion Rate (RCRATE)

Simulation Parameters:

- $T_{dt} = T_a = 6$ months (Maximum period for competency recovery).
- T_{sl} (Skill Loss Rate Averaging Time) is varied from 0 to 30 months.
- $T_s = T_{rd} = T_{td} = 3$ months

Simulation result is shown in Figure 5.20.

Inference:

- Higher T_{sl} values slightly decrease the time for rise of the Recruitment Completion Rate. Higher T_{sl} values also indicate that the rate of people being recruited is high. The settling time for higher values of T_{sl} is low.
- The system takes 75 months to settle, i.e., the system is able to recover from the fluctuations only after a duration of 75 months.
- The smaller the T_{sl} values, slightly higher the time taken for the Recruitment Completion Rate to rise. Smaller T_{sl} values also indicate that the rate of people being recruited is decreased. The settling time for lower values of T_{sl} is high.

For better results, the optimum value of T_{sl} should be ranging from a minimum of 3 to a maximum of 10 months.

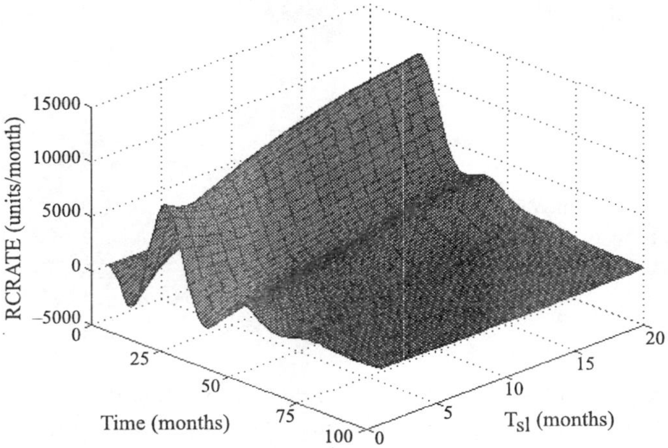

Figure 5.20: Response of RCRATE for T_{sl}.

(b) Influence on Training Completion Rate (TCRATE)

Simulation Parameters:

- $T_{dt} = T_a = 6$ months (Maximum period for competency recovery).
- T_{sl} (Skill Loss Rate Averaging Time) is varied from 0 to 30 months.
- $T_s = T_{rd} = T_{td} = 3$ months

Simulation result is shown in Figure 5.21.

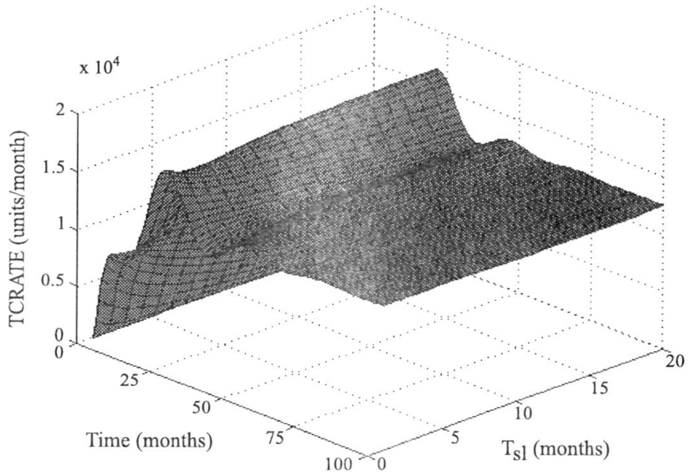

Figure 5.21: Response of TCRATE for T_{sl}.

Inference:

- Higher T_{sl} values slightly decrease the time for rise of Training Completion Rate. Higher T_{sl} values also indicate that the rate of people being trained is high. Settling time for higher values of T_{sl} is low.
- The system takes 60 months to settle, i.e., the system is able to recover from the fluctuations only after duration of 60 months.
- The smaller the T_{sl} values, the slightly higher will be the time taken for the Training Completion Rate to rise. Smaller T_{sl} values also indicate that the rate of people being trained is low. The settling time for lower values of T_{sl} is high.

For better results, the optimum value of T_{sl} should be ranging from a minimum of 3 to a maximum of 10 months.

(c) Influence on Actual Level of Competence Absorbed (ALCA)

Simulation Parameters:

- $T_{dt} = T_a = 6$ months (Maximum period for competency recovery).
- T_{sl} (Skill Loss Rate Averaging Time) is varied from 0 to 30 months.
- $T_s = T_{rd} = T_{td} = 3$ months

Simulation result is shown in Figure 5.22.

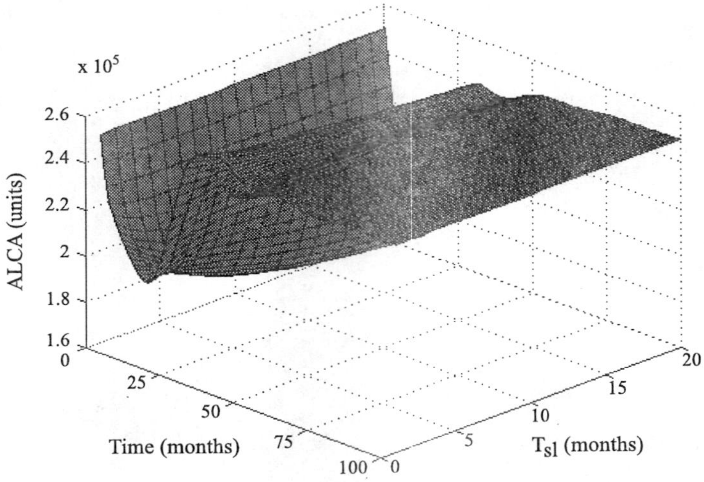

Figure 5.22: Response of ALCA for T_{sl}.

Inference:

- The larger the T_{sl} values, the larger the competence pool drop, indicating that the system is unable to recover from the competency shortages over a period of time.
- The competence pool takes 50 months to settle, i.e., the system is able to recover from the competency shortages only after a duration of about 50 months for higher values of T_{sl}.
- Lower values of T_{sl} indicate a smooth rise in the competence level for values of T_{sl}. This smooth rise is followed by fluctuations, after which the system response remains constant.

For better results, optimum value of T_{sl} should be ranging from a minimum of 3 to maximum of 8 months.

Comparative Analysis of Simulation Results

The comparative analysis of Engineering Competence Pool Model on Recruitment completion rate (RCRATE), Training completion rate (TCRATE) and Actual level of competence absorbed (ALCA) reveals the fact that all the six policy parameters chosen have considerable influence on the system dynamics of the model. The performance indices provide information on the optimum setting of policy parameters (Table 5.1).

- For better results in RCRATE and TCRATE, the time over which the competence gap is to be recovered (T_r) should be about 3 to 6 months. This also ensures a smooth rise in ALCA.
- The competence training delay time (T_{dt}) should be as short as possible, the most ideal being about 2 to 3 months, w.r.t. RCRATE and TCRATE, which also ensures smooth ALCA.
- The average time to determine the forecast competence loss rate (T_a) should not extend for more than 4 months (< 8 months) w.r.t. RCRATE, and 1 to 3 months w.r.t. TCRATE, but with reference to the ALCA the ideal duration would be the first 2 months.
- The skill gap recovery rate (T_s) has a significant influence on recruitment as the first few months may result in aggressive hiring and firing when there is over supply of engineers. Hence, in 4-6 months the fluctuations may settle down and peak RCRATE & TCRATE may be ensured, which may also provide a smooth ALCA.

Table 5.1: Performance Indices of Engineering Competence Model

Performance Indices		*Policy Parameters* T_r	T_{dt}	T_a	T_s	T_{td}	T_{sl}
Recruitment Completion Rate [RCRATE]	Rise time	Increasing T_r increases the time for rise	Increasing T_{dt} increases the time for rise	Increasing T_a increases the time for rise slightly	Increasing T_s increases the time for rise considerably	Increasing T_{td} increases the time for rise slightly	Increase in T_{sl} decreases the time for rise
	Peak Overshoot	Increasing T_r decreases the peak value	Increasing T_{dt} increases peak value, followed by a slight decrease	Increasing T_a decreases the peak value gradually	Increasing T_s decreases the peak value considerably	Increasing T_{td} increases the peak value gradually	Increase in T_{sl} decreases the peak value for T_s<2, followed by a increase
	Recruitment Completion Rate	Increasing T_r decreases the RCRATE	Increasing T_{dt} first increases then decreases the RCRATE	Increasing T_a decreases the RCRATE	Increasing T_s decreases the RCRATE gradually	Increasing T_{td} increases the RCRATE gradually	Increase in T_{sl} decreases the RCRATE initially followed by an increase
	Settling time	60 months [Fig. 5.5]	60 months [Fig. 5.8]	60 months [Fig. 5.11]	38 months for T_s ≥ [Fig. 5.14]	480 months [Fig. 5.17]	75 months [Fig. 5.20]
Training Completion Rate [TCRATE]	Rise time	Increasing T_r increases the time for rise	Increasing T_{dt}, increases the time for rise	Increasing T_a increases the time for rise slightly	Increasing T_s increases the time for rise	Increasing T_{td} increases the time for rise slightly	Increase in T_{sl} decreases the time for rise
	Peak Overshoot	Increasing T_r decreases the peak value	Increasing T_{dt} decreases the peak value	Increasing T_a decreases the peak value	Increasing T_s decreases the peak value	Increasing T_{td} increases the peak value gradually	Increase in T_{sl} decreases the peak value for T_s>2, followed by an increase
	Training Completion Rate	Increasing T_r decreases the TCRATE	Increasing T_{dt} decreases the TCRATE slightly	Increasing T_a decreases the TCRATE slightly	Increasing T_s decreases the TCRATE gradually	Increasing T_{td} increases the TCRATE	Increase in T_{sl} decreases the TCRATE followed by a increase
	Settling time	65 months [Fig. 5.6]	65 months [Fig. 5.9]	65 months [Fig. 5.12]	30 months for T_s>3 [Fig. 5.15]	80 months [Fig. 5.18]	60 months [Fig. 5.21]
Actual Level of Competence Absorbed [ALCA]	Competence pool drop	Increasing T_r increases the competence pool drop	Increasing T_{dt} increases the competence pool drop	Increasing T_a increases the competence pool drop	Increasing T_s increases the competence pool drop slightly	Increasing T_{td}, increases the competence pool drop	Increase in T_{sl} increases the competence pool drop
	Duration of competence pool deficit	Increasing T_r increases the settling time	Increasing T_{dt} increases the settling time	Increasing T_a increases the settling time	Increase in T_s increases the settling time for T_s>2	Increasing T_{td} increases the settling time	Increase in T_{sl} increases the settling time
	Competence pool overshoot	Increasing T_r decreases the peak value	Increasing T_{dt} decreases the peak value	Increasing T_a decreases the peak value	Increasing T_s decreases the peak value	Increasing T_{td} increases the peak value	Increase in T_{sl} decreases the peak value
	Settling time	60 months [Fig. 5.7]	65 months [Fig. 5.10]	70 months [Fig. 5.13]	78 months [Fig. 5.16]	90 months [Fig. 5.19]	50 months [Fig. 5.22]

- The industrial Training Delay Time (T_{td}) must be kept to a minimum and should not exceed 2-3 months for RCRATE & TCRATE, during which there would also be quick ALCA.
- Finally, the Skill Loss Rate Averaging Time (T_{sl}) should be between 3-6 months for the best results of RCRATE and TCRATE. This would also ensure smooth skill absorption ALCA.

From the superimposition of the simulation results (Figs. 5.23, 5.24 and 5.25), it can be observed that the Skill Gap Recovery Rate (T_s) is the most critical policy parameter because, higher T_s values:

- take more time to settle,
- take longer time for rise,
- result in lower peak value,

in comparison to the other policy parameters.

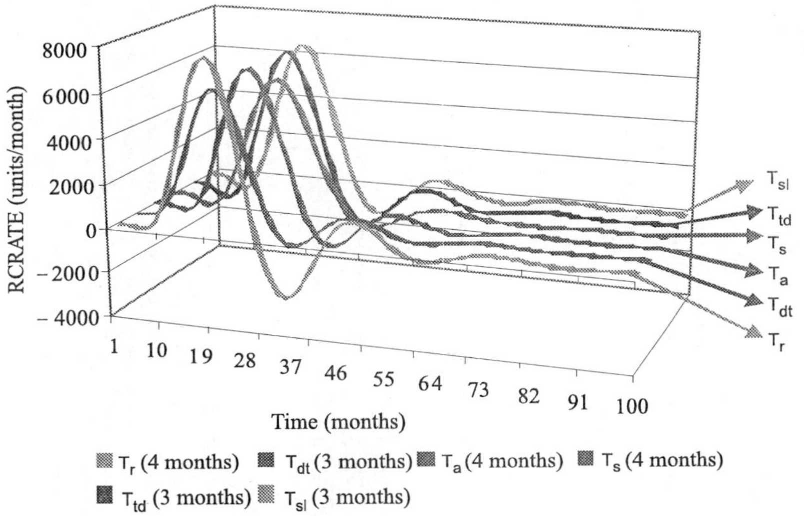

Figure 5.23: Superimposition of RCRATE in PECOPM.

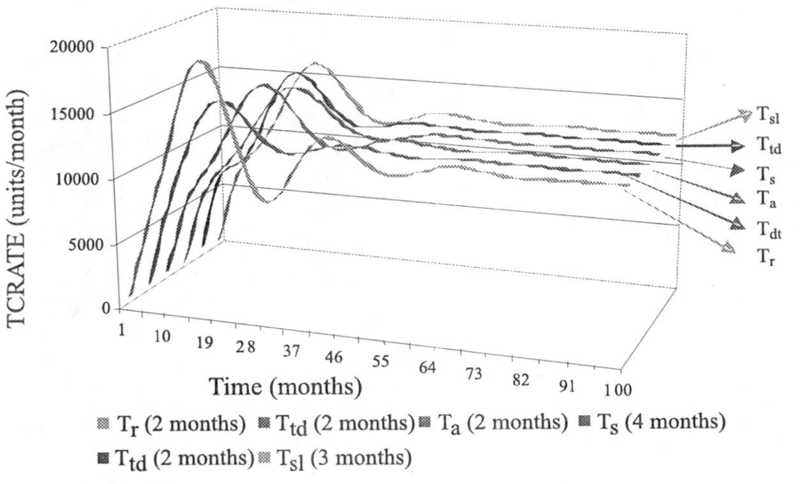

Figure 5.24: Superimposition of TCRATE in PECOPM.

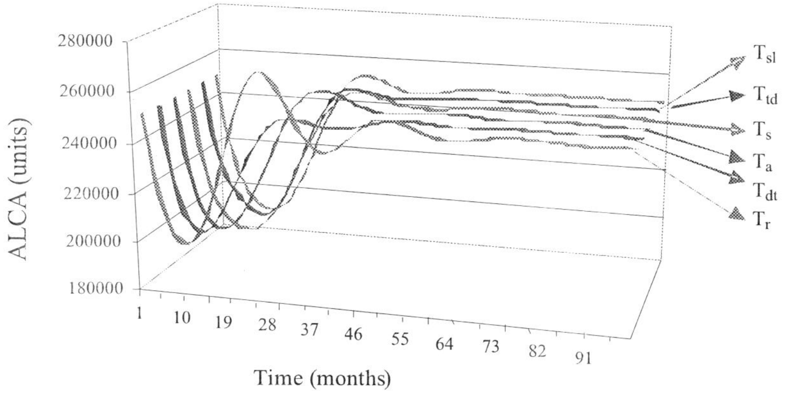

T_r (2 months) T_{dt} (2 months) T_a (4 months) T_s (4 months)
T_{td} (2 months) T_{sl} (3 months)

Figure 5.25: Superimposition of ALCA in PECOPM.

5.3.6 Model Validation

Validation has been one of the most elusive of the unresolved problems of social systems modelling (Mohapatra, 1987), be it simulation models in general, or system dynamics models in particular. Law & Kelton (1991) stated that there is no such thing as an absolutely valid model. Sterman (2000) argues that all models are, in a sense, wrong because there could always be a counter example to which the model does not conform to completely. Nevertheless, the perfect way to validate the system dynamics model would be to practically test it in real-life situations. However, it is the most expensive and time-consuming method, and hence, not very practicable. There is a separate chapter dedicated to Model Validation in which the same example has been explained in detail.

5.3.7 Policy Analysis and Improvement

The Performance Index table (Table 5.1) provides the necessary information regarding the influence of the time-based policy parameters on the dynamic system. Inferences are drawn to decide the most critical policy parameters, which have significant influence on the engineering competence pool model. Hence, a strong implication that can be drawn from this study is that the Skill Gap Recovery Rate is the most critical time-based policy parameter which needs to be closely monitored for the most efficient development and deployment of engineering competence of the nation.

5.4 LIMITATIONS OF THE STUDY

One of the major limitations of the System Dynamics approach is that the simulation results can only show the trends and not precise estimates. The model developed in this case study does not consider the influence of extraneous factors such as change in government policies, inflation rate, influence of globalization, privatization and localization, etc. which may also have some influence on the competence of engineers. It also involves managerial biases, which may creep in during the stage of conceptualization. Also, the validation of the model is relatively subjective. However, as the main focus of this study was on identifying the most critical time-based policy parameters, which significantly influence the engineering competence pool development and deployment, these limitations are not major impediments.

5.5 SUMMARY

In this chapter, a case study on the Engineering Competence Pool Development and Deployment has been undertaken. The standard methodology of System Dynamics has been depicted through a diagram and it has been adopted in solving the proposed problem. The reader is encouraged to go through the MATLAB® manual to gain greater knowledge on using the software to produce 3D-surface graphs, however, the focus here is to conceptualize the problem and decide on the variables of simulation and the interpretation of the simulation results. The Policy Implementation based on simulation results is given so that the reader would be able to apply the knowledge gained in a similar situation elsewhere. Finally, the limitations of this study have been highlighted which opens up new possibility of extending this problem with additional set of variables.

5.6 EXERCISE

Conduct a survey on the total number of Management graduates produced in the country and their absorption rate. Develop a System Dynamics model to study the delay in various stages from graduation to recruitment and produce simulations to identify the most critical time-based policy parameter which needs to be closely monitored. What would be your suggestion for policy implementation?

6

Insurance Sector Dynamics

Objectives

After going through this chapter you will be able to:

- Identify the variables in Insurance Sector Dynamics.
- Build a model to associate the variables that represent the Service quality gap in Insurance sector.
- Conduct simulation.
- Interpret the simulation results.

6.1 PROBLEM STATEMENT

The Regional Head of ABC Insurance Company has encountered a problem situation where her quarter yearly sales of insurance policies have declined drastically. The reason for this, according to her preliminary study, is the growing 'gap' between customer *perceptions* and *expectations*.

The Regional Head calls the Manager (QC) and assigns him the task of preparing a questionnaire to collect the feedback from their customers to measure the gap in the service quality perceptions and expectations. But before preparing the questionnaire, she instructs the manager to identify the antecedents of service quality perceptions and expectations and build a System Dynamics model to represent Insurance sector dynamics and conduct simulations to study the influence of these four factors on service quality gap. Presuming yourself to be the Manager (QC), undertake the assignment.

6.2 SOLUTION

6.2.1 Insurance Sector Dynamics

The causes tree diagram for 'service quality perception' and 'service quality expectation' (Sureshchandar et al., 2001; Parasuraman et al., 1988; Bitner, 1992; Kotler, 1973; Zeithaml et al., 1985) are given in Figures 6.1 and 6.2.

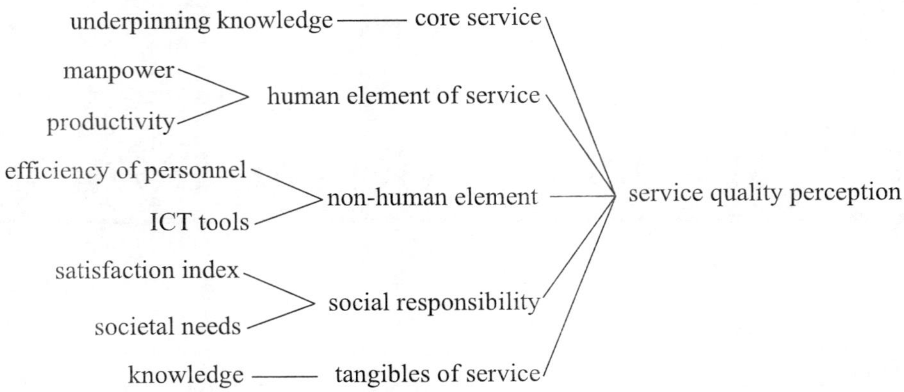

Figure 6.1: Causes Tree–Service Quality Perception.

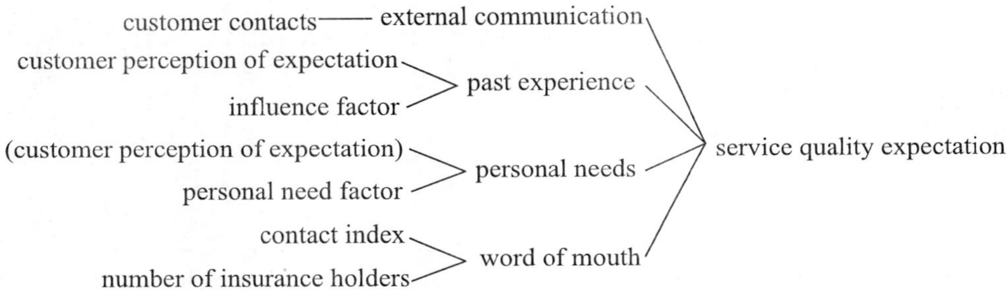

Figure 6.2: Causes Tree–Service Quality Expectation.

The causes tree diagram depicts the causal relation between the variables. It can be observed that service quality perception is influenced mainly by five factors, viz., core service, human element of service, non-human element, social responsibility, and tangibles of service. Similarly, service quality expectation of customer is formed based on 4 primary factors, viz., External communication, Past experience, Personal needs, and Word of mouth. These factors have been identified by a group of researchers on service quality as listed before. There may be other influencing factors which are beyond the scope of this book and the reader may undertake a serious research to identify some more factors and modify the basic model developed in this chapter.

6.2.2 Causal Diagram

Based on the causes tree diagram given in the previous section the Causal diagram may be drawn to give the fundamental understanding of the cause and effect relation between the variables holistically. Figure 6.3 depicts the Causal loop diagram of insurance sector dynamics.

6.2.3 Stock and Flow Diagram

The causal diagram provides the foundation to build the stock and flow diagram. The stock and flow diagram is shown in Figure 6.4. The governing equations are given in Appendix-II.

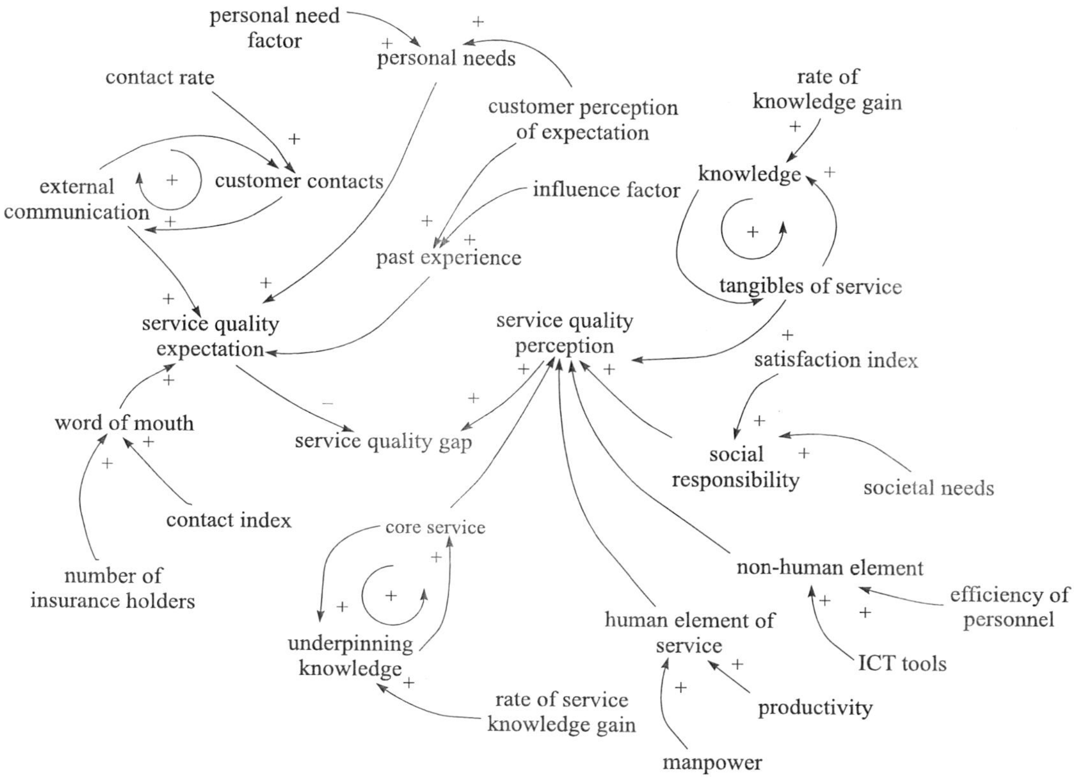

Figure 6.3: Insurance Sector Dynamics, Causal Loop Diagram.

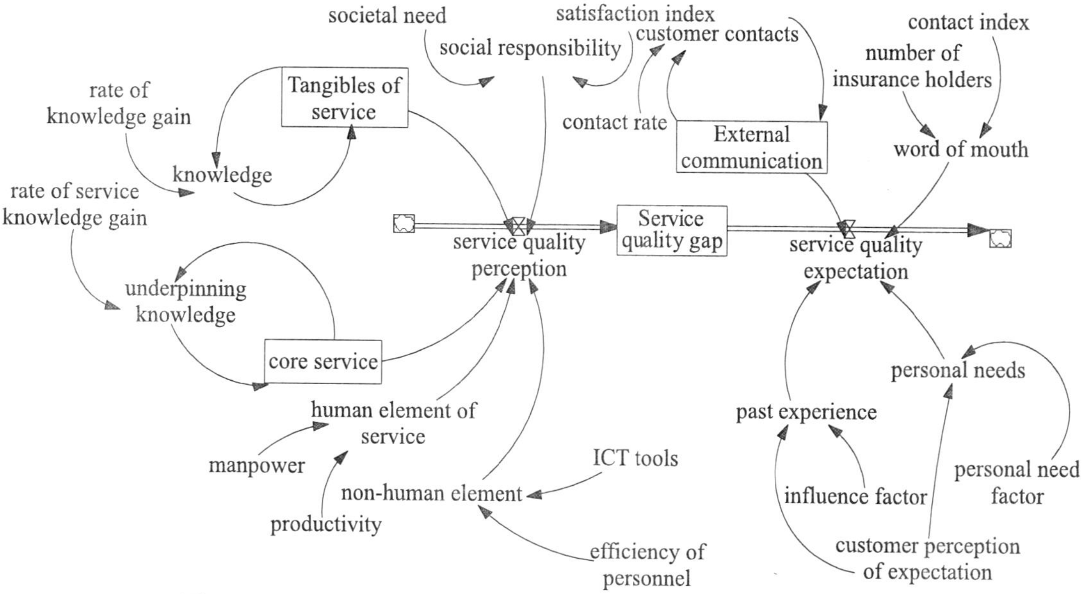

Figure 6.4: Insurance Sector Dynamics—Stock and Flow Diagram.

The stock and flow diagram fundamentally is based on the relation,

Service quality gap = Service quality perception - Service quality expectation

It is important to note that quantifying service quality on an absolute scale is a difficult proposition compared to measurable vectors such as force in newton, velocity in m/sec, etc. However, in System Dynamics it is always the trend and the pattern of variation that is of significance and not the accuracy of the numbers and absolute units. But in the equations used in the model, the units of LHS must balance with that on the RHS. Hence, we may quantify that a given number of policies sold, or a given amount of time spent on customers, or a perceptional average of some number on Likert's scale of measurement represents a 'unit' of service quality measurement.

It can be observed from the stock and flow diagram that the rate variables, viz., service quality perception and service quality expectation are influenced by the parameters as related in the causal diagram.

6.2.4 Simulation

The primary objective of this exercise is to study the critical factors which have influence on the quality of service provided in an Insurance sector. Following are the critical parameters whose influence on system performance has been studied in order to make policy implementations.

1. *Past Experience*

Past experience is the customer's previous exposure to service that is relevant to the same company or some other insurance company. Past experience is one of the forces that shape predictions and desires. The past experience of a customer on issues such as recommendation on right policy, bargain on premium, experiences in dealing with insurance officials, claim experiences, time period to recover claim settlement, etc. play a very important role in forming perceptions. When an employee who has maintained the record of such past experience of customers leaves the organization, he carries away all the explicit, implicit and tacit knowledge with him. Also, the past experience of the customer, on the other hand, will be a key parameter that will make him to repurchase insurance policies or not.

The simulation results indicate that the 'service quality gap' is influenced very little by the increase in the 'customer expectation of performance' to start with, however, the variation becomes significant over an extended period of time. In this simulation, the variables (Figure 6.5): ceper1 to ceper5, refer to the 'customer expectation', which is based on past experience, increased from 1000 units to 3000 units in steps of 500. It can be observed that the lower the customer expectation, the higher will be the service quality gap. It is interesting to note that service quality gap will be minimum at the starting stage (even a negative trend is observed at the start, i.e., expectation is higher than the perceived quality), and over a period of time the perceptions improve gradually.

2. *Personal Needs*

The customers' personal needs can be of physical, social, psychological, and functional like: fulfil obligations, learn my business and work with me, protect me from catastrophe, provide prompt service, etc. While conducting the simulation runs, a given unit of *past experience*, the *personal needs*, which in turn is influenced by the *personal need factor* was varied from 0.2 to 0.6 in steps

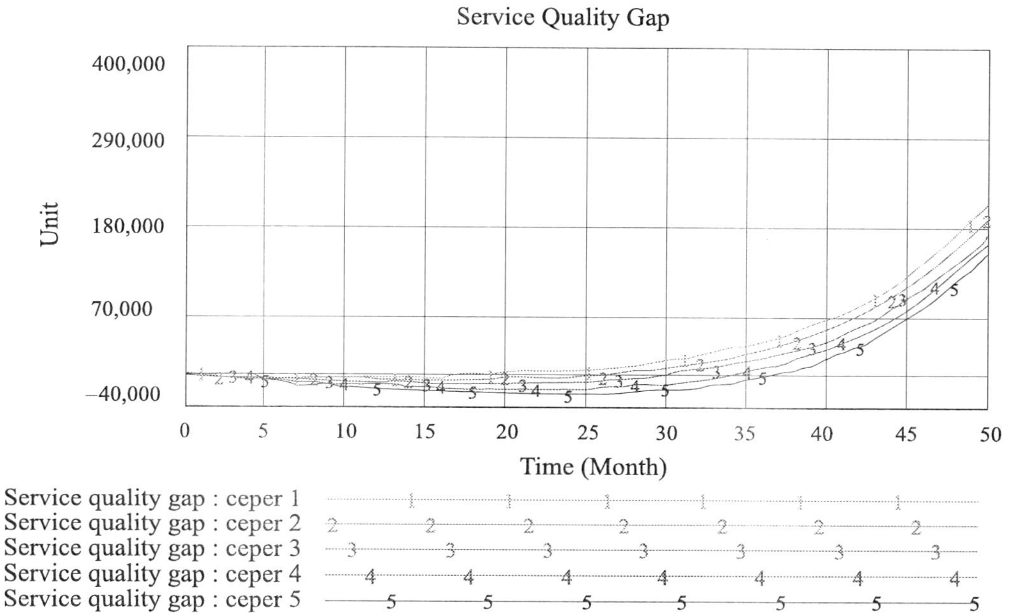

Figure 6.5: Past Experience versus Service Quality Gap.

of 0.1 (Figure 6.6). With the *past experience* coupled with increasing personal needs of the customer, the *service quality expectation* dominance may be observed to start with, but eventually, their *service quality perception* improves over a period of time. It is important to note that the higher the *personal needs*, the longer will be the service quality recovery period.

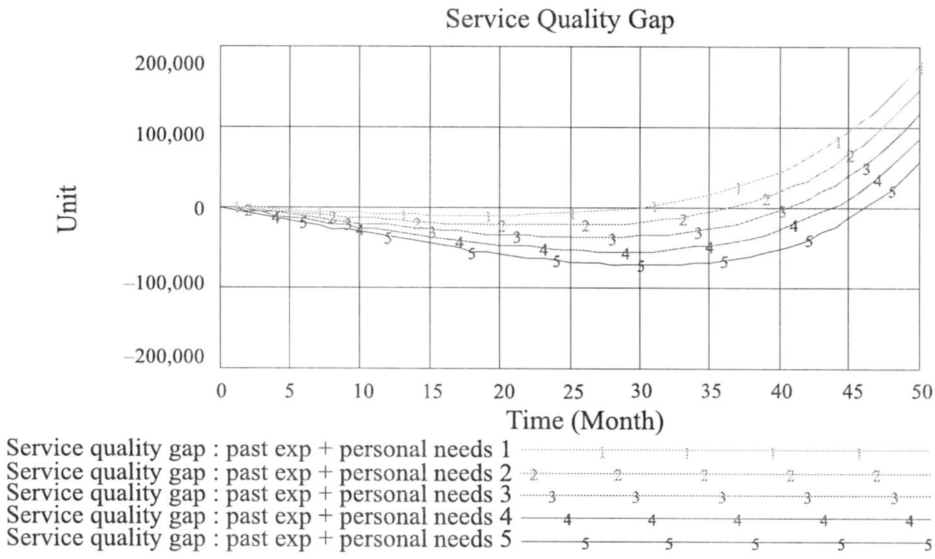

Figure 6.6: Service Quality Gap versus Personal Needs.

3. External Communication

External communication is mainly through the customer contact external to the organization and is influenced primarily by the *contact rate* of the customers. In this simulation, the *contact rate* was varied from 0.1 to 0.2 in steps of 0.025 and its influence on *service quality gap* was studied (Figure 6.7). It can be observed that by and large there will be no change in the *service quality gap* but eventually after an extended period of time (> 30 months) the sector fails to meet the customer expectation.

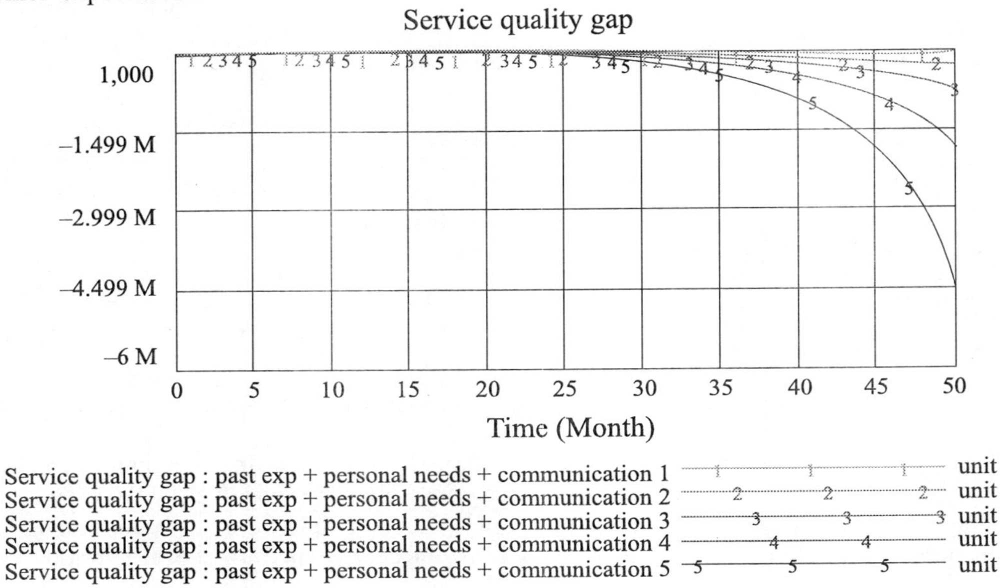

Figure 6.7: Service Quality Gap versus External Communication.

4. Word of Mouth

In this case, the influence of *word of mouth* on the *service quality* is studied. The *contact index* will decide the number of contacts made by the customers, and hence, it was varied from 0.2 to 0.8 to study the influence (Figure 6.8). It can be observed that with the increase in the *word of mouth* mode of communication, there is a steady decrease in the *service quality gap*, because the *service quality expectation* increases continuously. However, after reaching particular duration of time (> 40 months), the system recovers, and thereafter, *service quality perceptions* will take over.

6.2.5 Policy Implementation

Having studied the influence of various factors on service quality, the manager (QC) makes the following recommendations to the Regional Head for policy implementation:

1. Enhancing customer expectation.
2. Creating awareness to increase the customer needs.
3. Promoting internal communication instead of external communication.
4. Delighting the customers to promote word of mouth.

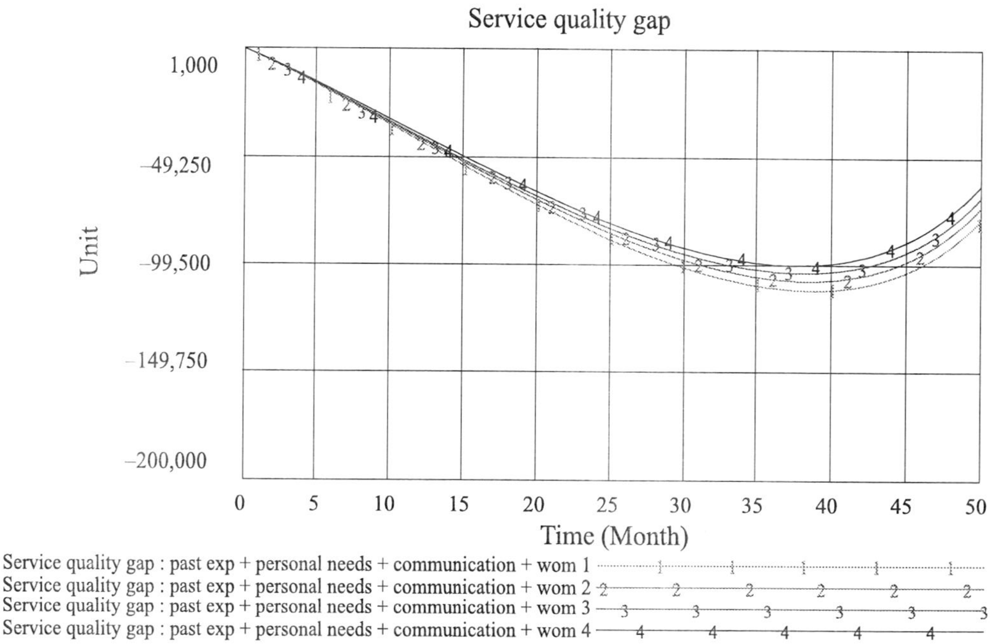

Figure 6.8: Service Quality Gap versus Word of Mouth.

A closer look at the above implications calls for the transformation of the Insurance sector into a learning organization through the introduction of a Knowledge Management System (KMS) because all the above implications in one way or the other are dependent on the speed and agility of the service provider to assimilate, validate, store, distribute and apply the knowledge pertaining to: customers, product, services, and the progress made by science and technology. So the main implication of the study is to develop a KMS that could act as a storage place of all the knowledge pertaining to insurance sector w.r.t. the types of policies, nature of claims, sharing of best practices, lessons learned, service add-ons, etc. On this common portal all the stakeholders can access the relevant information and be in touch with the recent happenings.

6.3 SUMMARY

This chapter takes an example of Insurance sector and identifies the key variables associated in the system dynamics. It deals with the building of causal loop diagram and stock and flow diagram and thus the model building. The simulation results of critical factors have been explained and the policy implementation has been made so as to give an understanding of how System Dynamics can be used as a tool in problem solving.

6.4 EXERCISE

Conduct a literature review on service quality measurement and for a service sector such as bank, travel and tourism, hospitality, health care, etc. identify the key factors that influence service quality and develop a System Dynamics model to simulate the influence and suggest Policy Implementations.

7

Manufacturing Supply Chain Dynamics

Objectives

After going through this chapter you will be able to:

- Identify the key variables in Manufacturing Supply Chain.
- Build a model to associate the variables that represent the Supply Chain Management in a typical Manufacturing Sector.
- Conduct simulation on Supply Chain Dynamics.
- Interpret the simulation results and make suggestions for Policy Implementation.

7.1 INTRODUCTION

Manufacturing is full of activities and events starting from raw material procurement and ending with distribution of product. Supply Chain Management has become an important component of Operations Management because it links customers and suppliers at each stage of operation from raw material procurement to delivering of the finished goods. In essence, it is concerned with how to effectively manage the relationship between suppliers, manufacturers and customers. Supply chain inventories are prone to fluctuations and instability. Even small variations in end item demand, create oscillations that amplify throughout the chain, which is called the 'bull-whip effect' (Lee et al., 1997; Sterman 2000). The first published work on supply chain management using system dynamics was way back in 1957. One of the most accepted Supply Chain Model (SCM) over the years is the Forrester model (Forrester, 1958), which has been very highly appreciated, as well received constructive criticism from supply chain researchers, thus leading to the improvement of the supply chain model on a continual basis. It has to be understood that as supply chain dynamics is a complex process, particularly in this globalized era where suppliers and distributors are spread across the world, the new parameters keep adding to the model and that makes the SCM highly dynamic in nature.

The model described in this chapter is based on the Sterman's manufacturing firm model built by Panov and Shiryaev (2003). For the sake of analysis, we take the supply chain of a small-scale aluminium utensil manufacturing industry. The model includes the process of goods production, transportation to distribution inventory, storage and waste in inventory, and distribution to the customers.

As stated before, the increased complexity in the Supply Chain Management (SCM) over the past years has necessitated the introduction of several parameters, which influence the efficiency of SCM. The lead times have become shorter and the supplier base has widened due to increased competition. In this chapter, the focus is on the study of the influence of changes in production rate, delay, waste time, and demand.

7.2 CONSTRUCTION OF SUPPLY CHAIN MODEL

7.2.1 The Dependence Diagram

The Dependence diagram indicates the causal relation between an independent variable and dependent variable. There can be more than one independent or dependent variables, and hence, sometimes it is even referred to as causal tree diagram. Before developing the SCM model, it is very important to have a clear understanding of all the associated variables and their nature of interdependencies. The Figure 7.1 depicts the Dependence diagram of SCM variables. It can be observed that adaption to production start rate and desired production rate influence desired production start rate. Further, available inventory and desired production start rate have a combined influence on production start rate. Similarly, other interrelations have been shown in Figure 7.1.

7.2.2 Causal Loop Diagram

The model is constructed based on the operations of an aluminium utensils manufacturing industry. Firstly, the factors influencing the production start rate is considered. One of the factors influencing the production start rate is the available inventory or available stock in the industry. The *available stock* is to be subtracted from *desired production start rate* to get the production start rate.

The *desired production rate* is influenced by several factors such as productivity of machines and men, capacity of each machine and number of men and machine available. The condition to be satisfied here is that each machine has a worker for working on it or each worker works on only one machine. To satisfy this condition, minimization function has to be used. The product of these factors is added with *adaptation to production start rate* and *expected inventory output* to get the desired production rate.

Delay can be further explained by adding factors such as delay due to *shipments from suppliers* and *shipments from maintenance.*

Demand can be assumed as the product of *forecasts* and *quality* of the product. Quality may be considered to be *actual quality* of the product multiplied by the *perception* or *expected quality* of the product. The perception of quality can be a number ranging from 0 to 1 for the sake of analysis, zero representing no quality and one representing the ultimate quality, both being purely theoretical. Hence, the *quality perceptive index* is varied from 0 to 1. Same scale is also used for the actual quality. The forecasting method used in this model is based on average sales for one year.

Waste time can be assumed as the sum of *rework time*, *idle time*, and *time lost in maintenance.* Idle time is assumed as the time when there is no production activity in progress. The initial conditions considered in this model are 1700 units of available inventory based on the data obtained from an aluminium utensil manufacturing company.

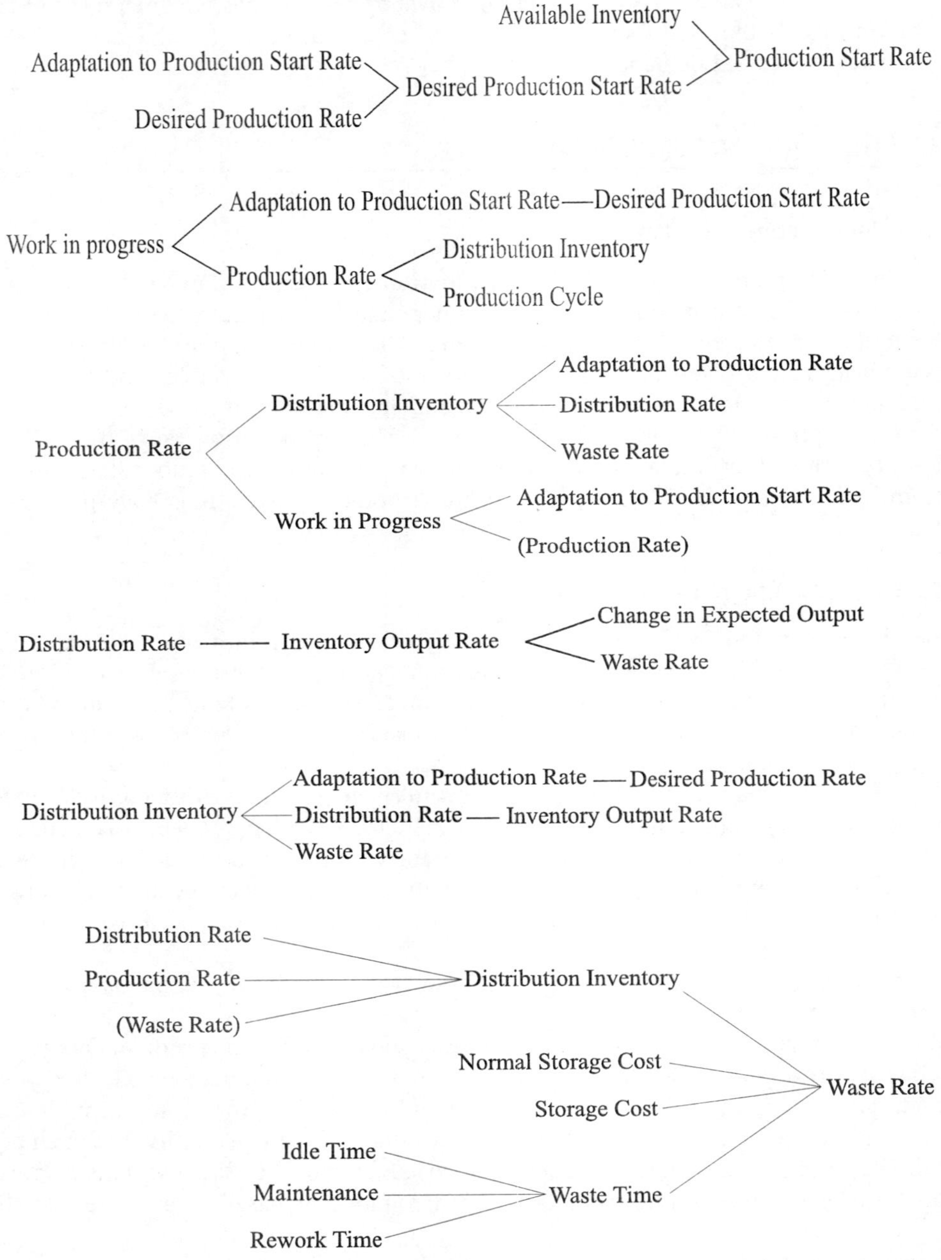

Figure 7.1: The Dependence Diagram of Expected Inventory Output.

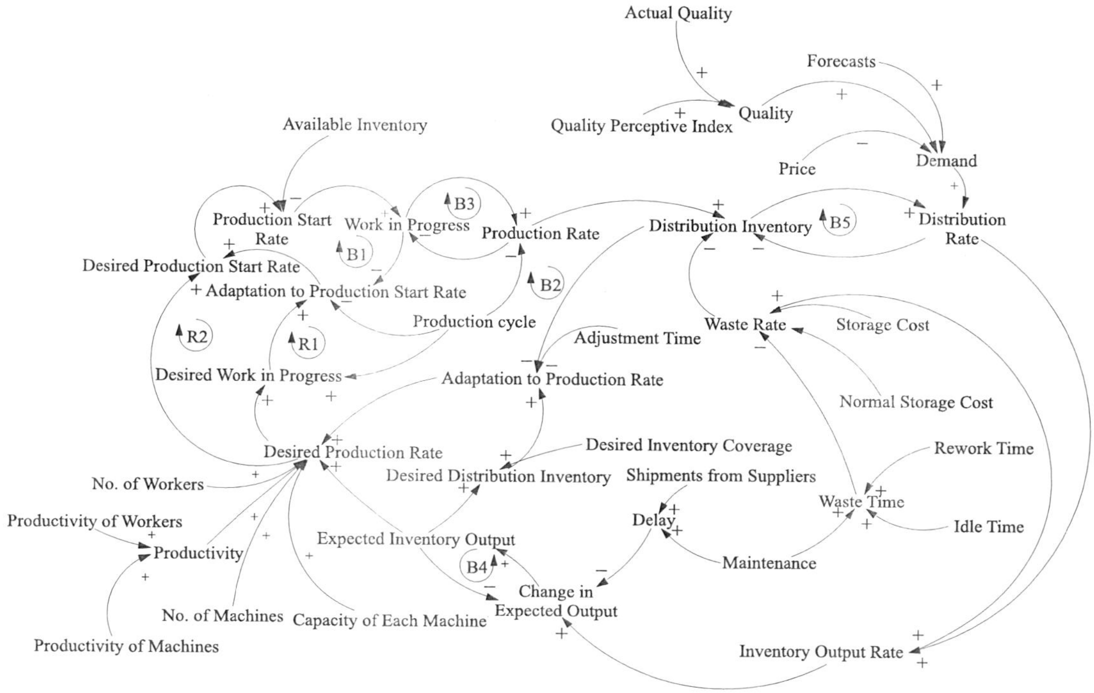

Figure 7.2: Causal Loop Diagram.

The causal loop diagram (Figure 7.2) shows the cause and effect relation between all the associated variables. It can be observed that there are five balancing loops and two reinforcing loops. The path of each loop is as given below:

R1 = Production Start Rate → Work in Progress → Production Rate → Distribution Inventory → Distribution Rate → Inventory Output Rate → Change in Expected Output → Expected Inventory Output → Desired Production Rate → Desired Production Start Rate

R2 = Production Start Rate → Work in Progress → Production Rate → Distribution Inventory → Distribution Rate → Inventory Output Rate → Change in Expected Output → Expected Inventory Output → Desired Production Rate → Desired Work in Progress → Adaptation to Production Start Rate → Desired Production Start Rate

B1 = Production Start Rate → Work in Progress → Adaptation to Production Start Rate → Desired Production Start Rate

B2 = Production Start Rate → Work in Progress → Production Rate → Distribution Inventory → Adaptation to Production Rate → Desired Production Rate → Desired Work in Progress → Adaptation to Production Start Rate → Desired Production Start Rate

B3 = Work in Progress → Production Rate

B4 = Expected Inventory Output → Change in Expected Output

B5 = Distribution Inventory → Distribution Rate.

7.2.3 Stock and Flow Diagram

The governing equations used in the model formulation are given in Appendix-III. The stock and flow diagram for the SCM is shown in Figure 7.3. The different stocks in the diagram are Work in Progress, Distribution Inventory and Expected Inventory Output. Work in progress is influenced by Production start rate, Production rate and Adaptation to production start rate. Distribution inventory is influenced by Production rate, Distribution rate and Waste rate. Distribution inventory influences Adaptation to production rate. Expected inventory output is influenced by Change in expected output. Expected inventory output influences Desired production rate and Desired distribution inventory.

7.3 SIMULATION RESULTS, ANALYSIS AND INFERENCES

The model was simulated for 60 runs and the influence on production start rate, work in progress, production rate, distribution inventory, distribution rate, and waste rate were studied.

Effect of Available Inventory: The graphs (Figs. 7.4 to 7.11) show the influence of available inventory on different factors. We can see that production start rate (Fig. 7.4), work in progress (Fig. 7.5), production rate (Fig. 7.6), distribution inventory (Fig. 7.8), distribution rate (Fig. 7.10), expected inventory output (Fig. 7.9), change in expected output (Fig. 7.11) and waste rate (Fig. 7.7) decrease with the increase in available inventory. It takes about 30 months to stabilize when inventories are maintained against about 50 months when normal inventory is maintained. The obvious reason is that having a larger safety stock reduces the production required to meet the demand.

Effect of Production Cycle Time: The graphs (7.12 to 7.19) show the influence of production cycle time on different factors. The simulation results depict that production start rate (Fig. 7.12) increases initially, and then deceases gradually and stabilizes with increase in production cycle time. The work in progress (Fig. 7.13) increases steeply to start with, and then, increases gradually with increase in production cycle time. Further, the production rate (Fig. 7.14) decreases with increase in production cycle time. Distribution rate (Fig. 7.16) remains constant for some time and then increases with increase in production cycle time. Distribution inventory (Fig. 7.15), waste rate (Fig. 7.17) decrease initially, and then, increase with the increase in production cycle time. Change in expected output (Fig. 7.18) decreases steeply initially, and then, increases steeply, after which, it gradually decreases and starts increasing slowly with increase in production cycle time. Expected inventory output (Fig. 7.19), on the contrary, increases initially, and then, decreases gradually and starts to increase slowly with increase in production cycle time. It has to be noted that in simulation study, more importance has to be given to the analysis of the trend than the numbers, such as 60 months, which is comparatively a larger time frame for a study. Hence, the key revelation of this study is the fact that Bullwhip effect is better controllable in a longer run of production cycle in comparison to the shorter one.

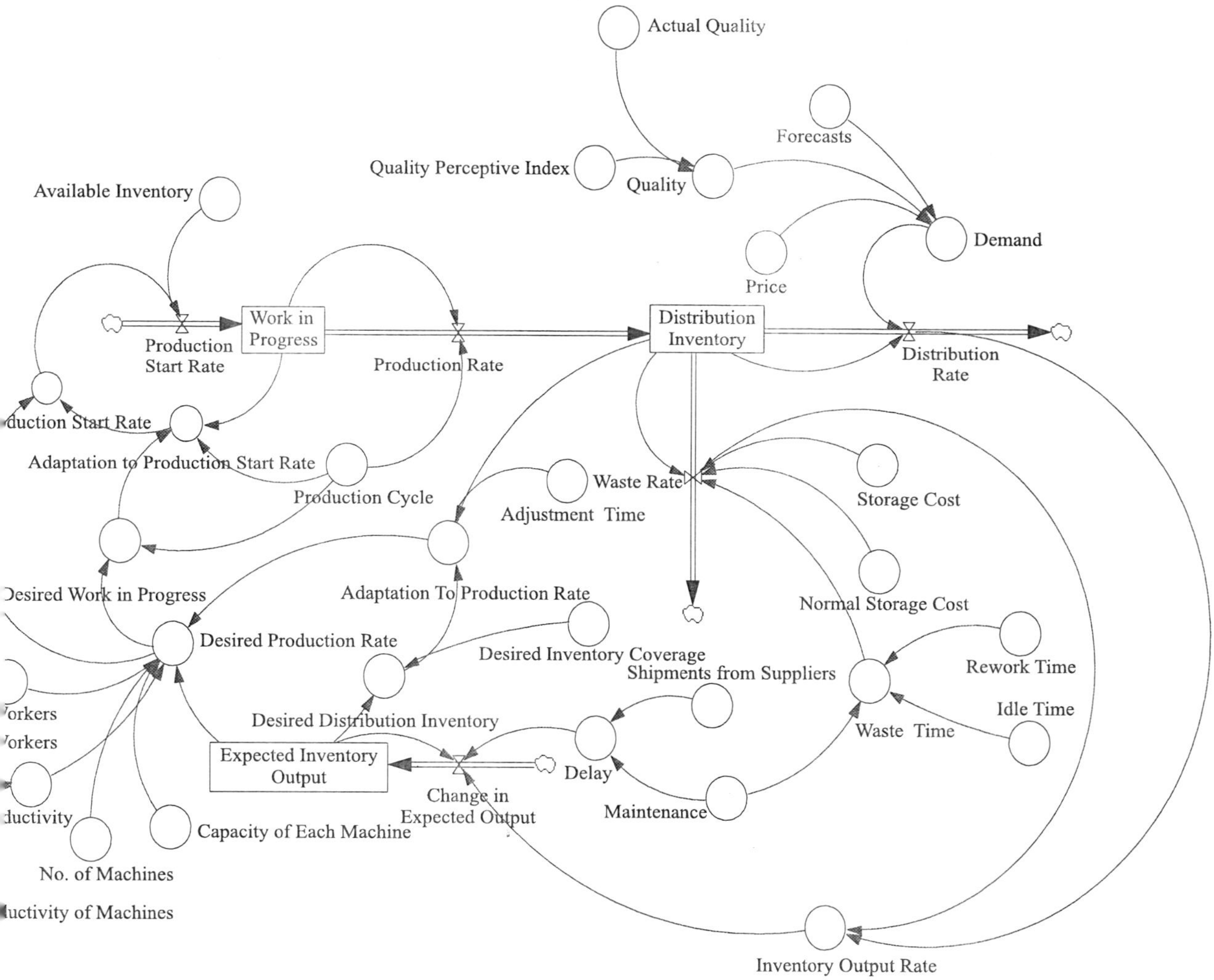

Figure 7.3: Revised Manufacturing Supply Chain Model with Feedback.

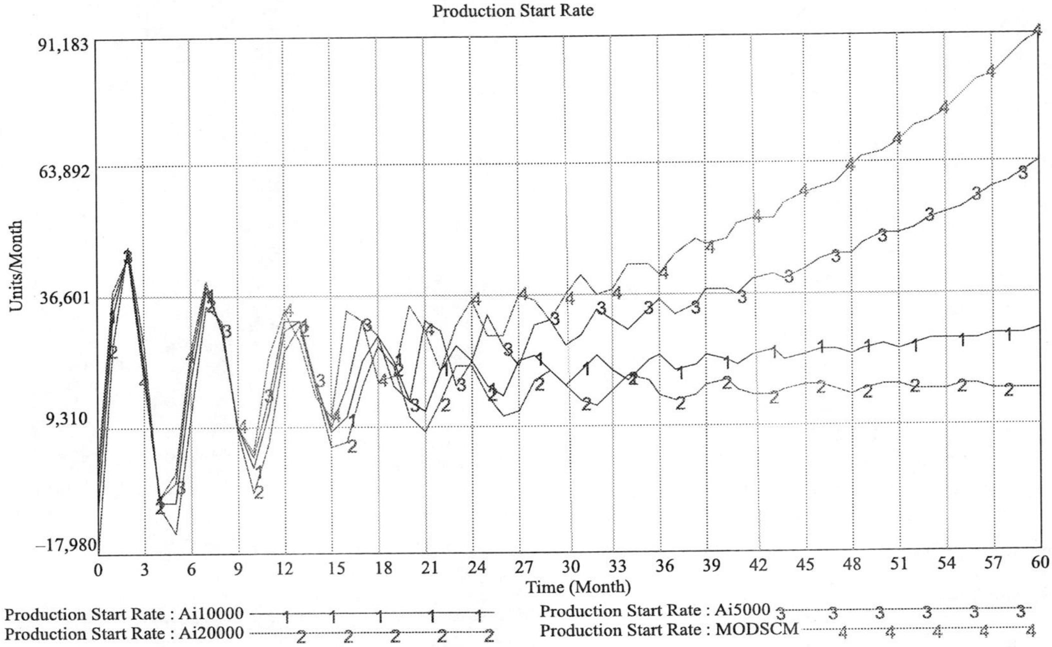

Figure 7.4: Effect of Available Inventory on Production Start Rate.

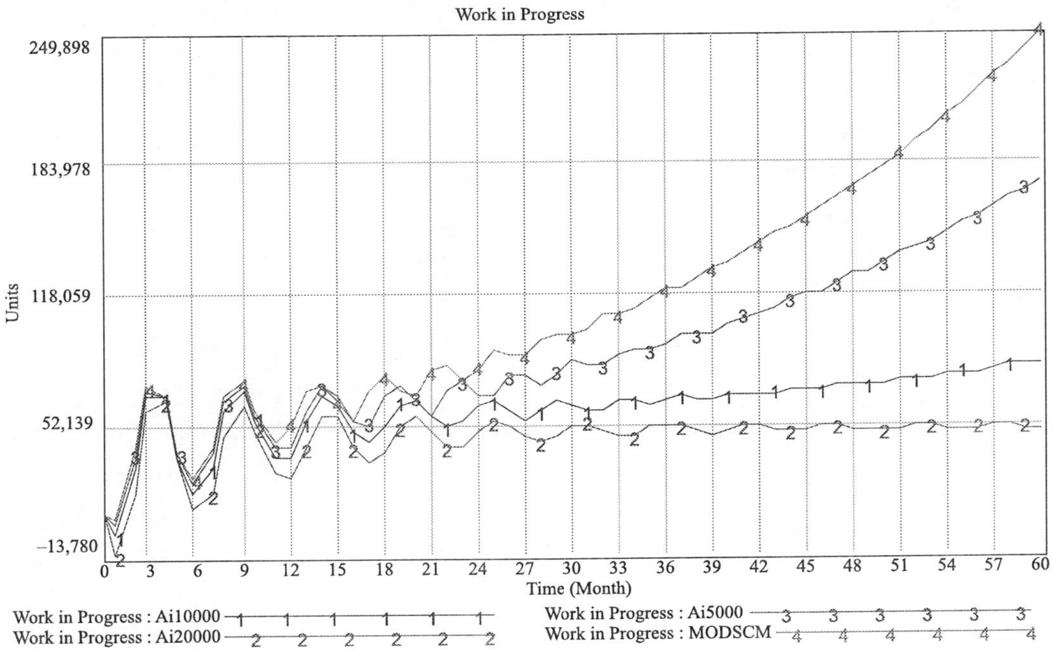

Figure 7.5: Effect of Available Inventory on Work in Progress.

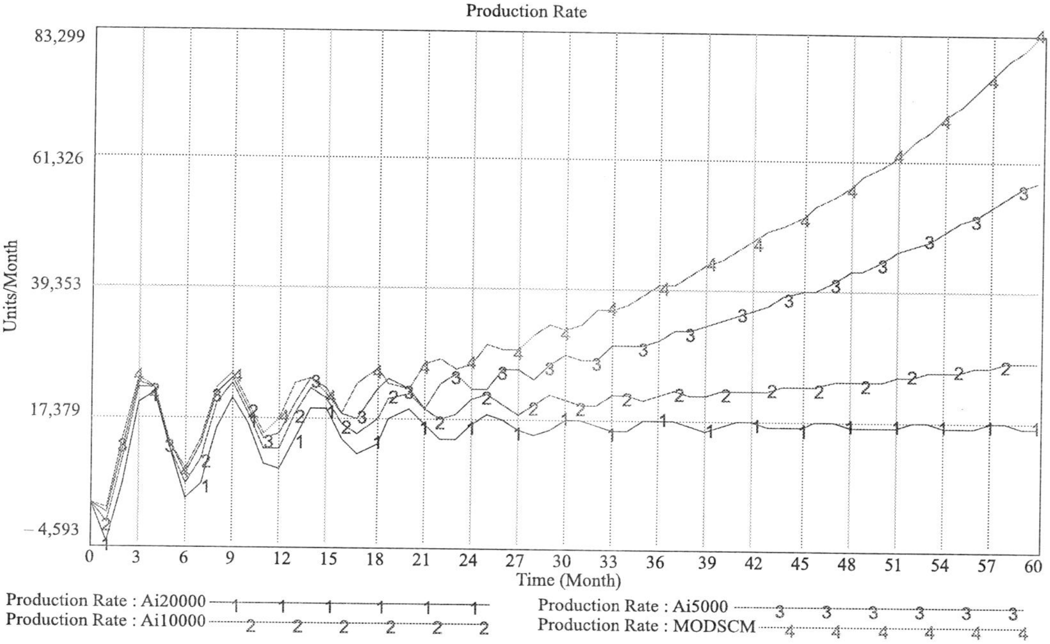

Figure 7.6: Effect of Available Inventory on Production Rate.

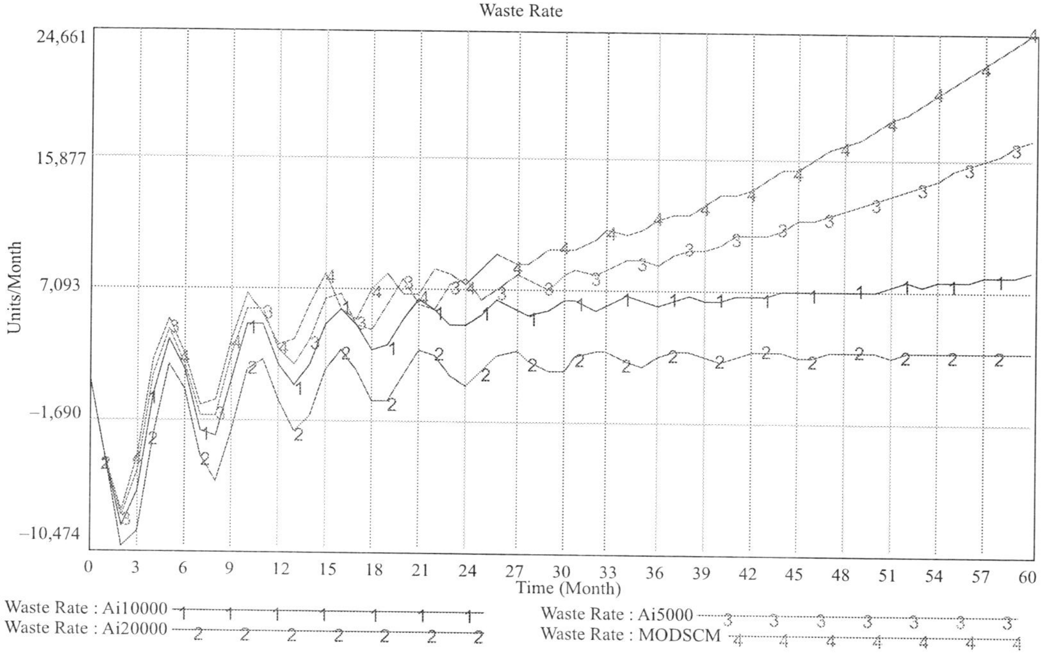

Figure 7.7: Effect of Available Inventory on Waste Rate.

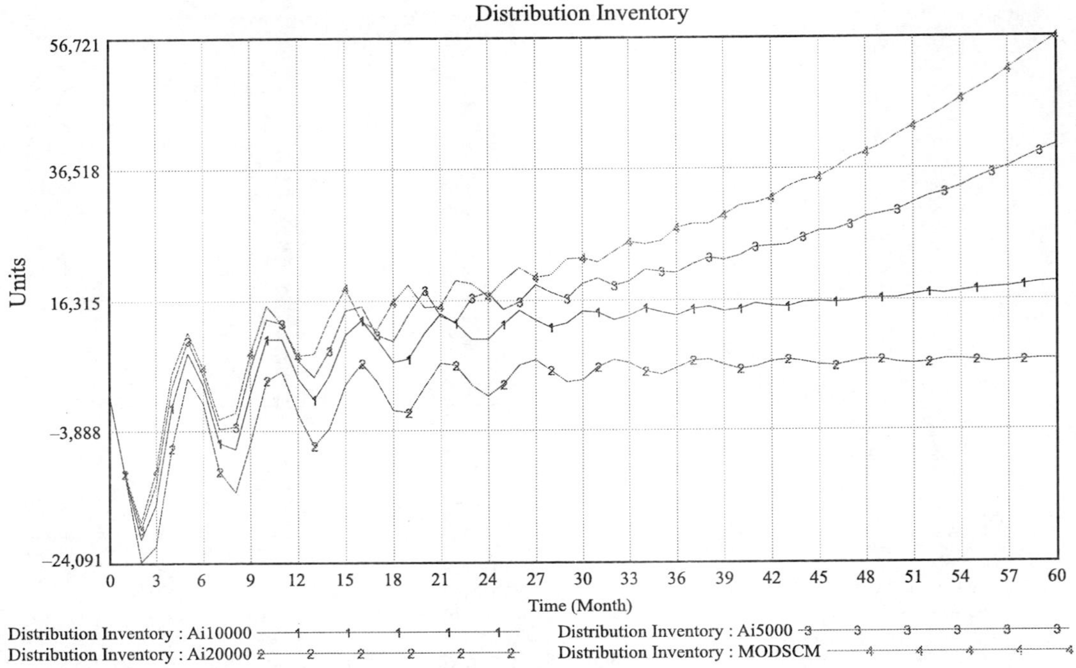

Figure 7.8: Effect of Available Inventory on Distribution Inventory.

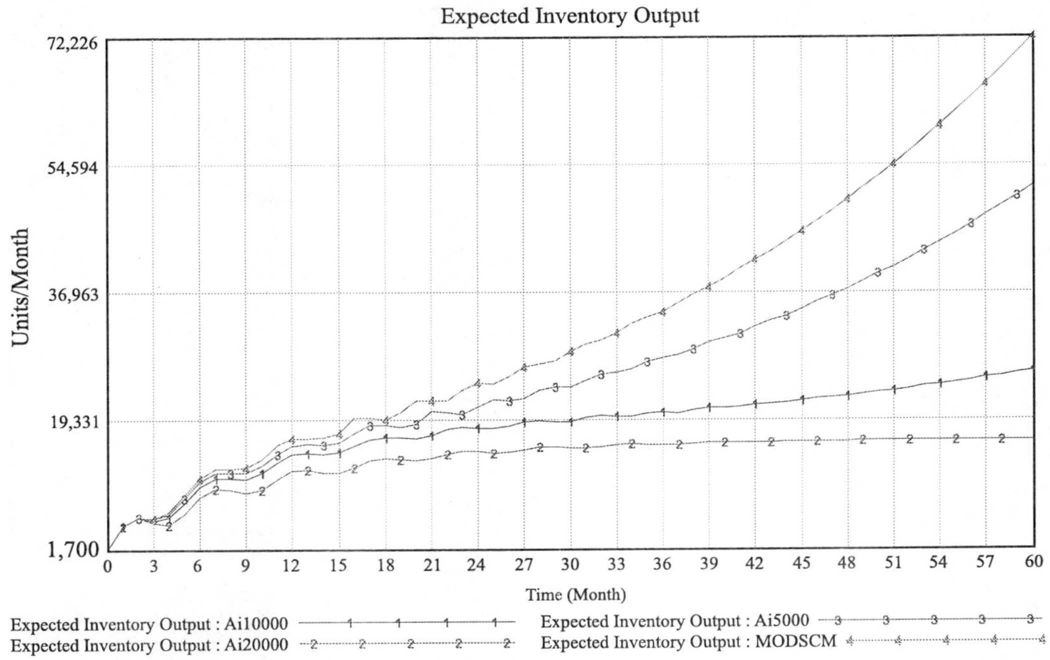

Figure 7.9: Effect of Available Inventory on Expected Inventory Output.

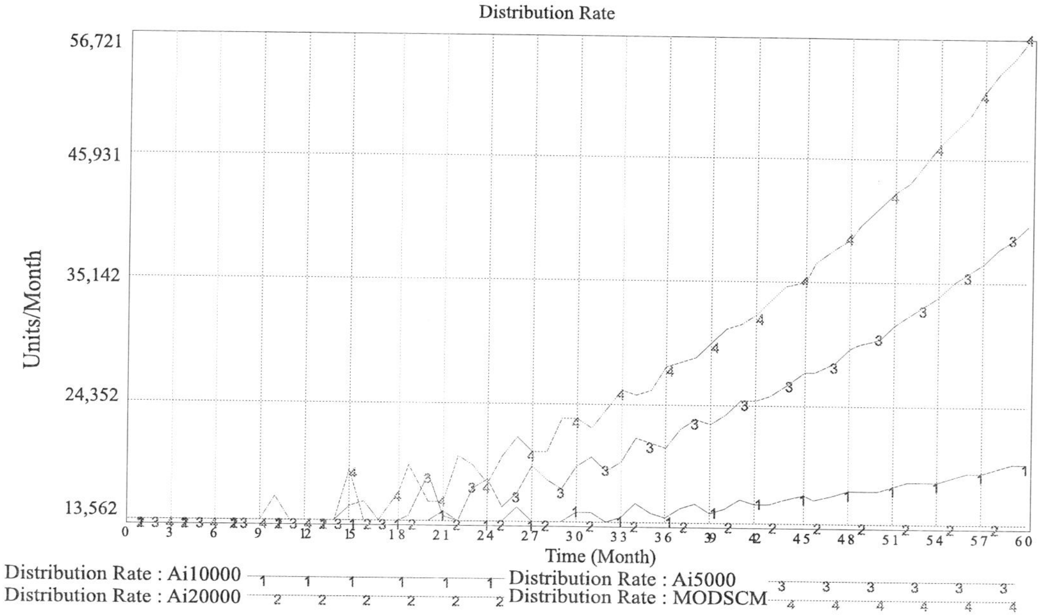

Figure 7.10: Effect of Available Inventory on Distribution Rate.

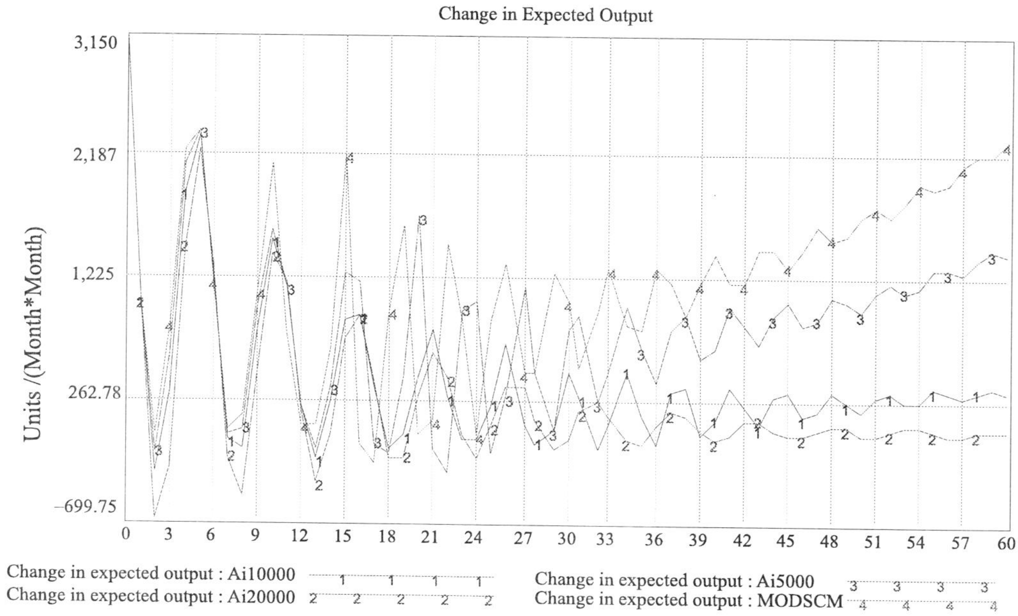

Figure 7.11: Effect of Available Inventory on Change in Expected Output.

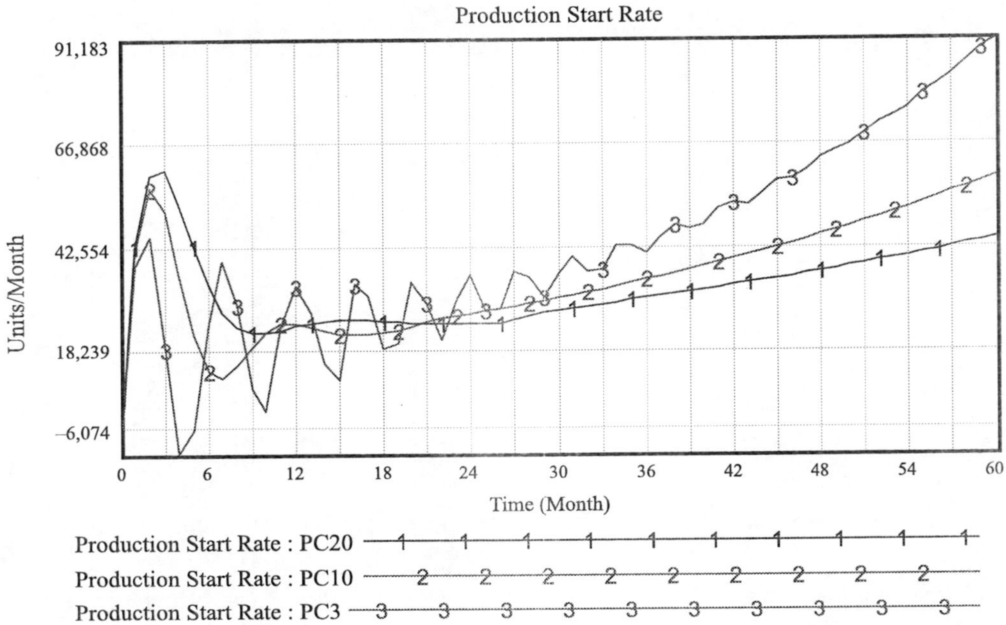

Figure 7.12: Effect of Production Cycle Time on Production Start Rate.

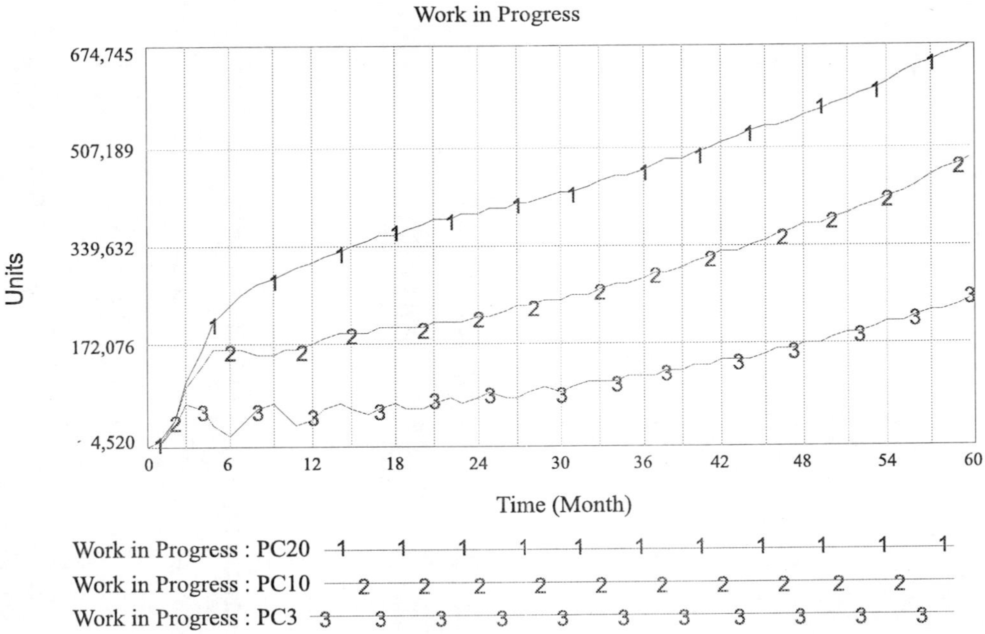

Figure 7.13: Effect of Production Cycle Time on Work in Progress.

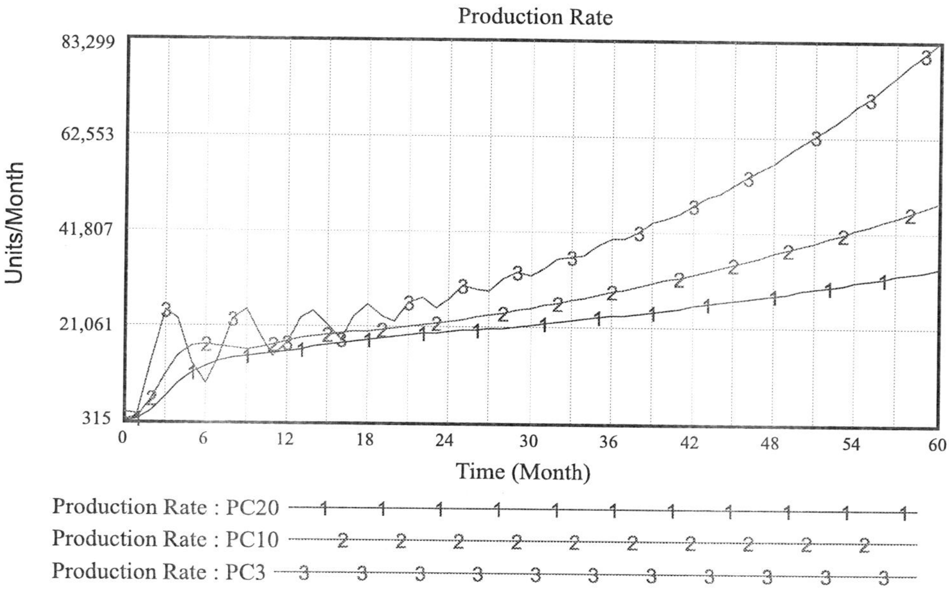

Figure 7.14: Effect of Production Cycle Time on Production Rate.

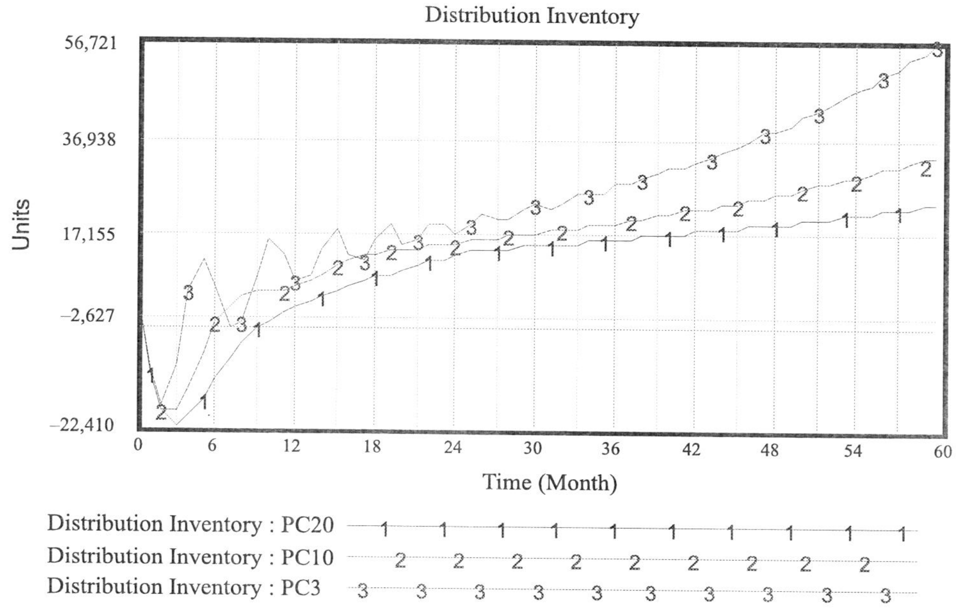

Figure 7.15: Effect of Production Cycle Time on Distribution Inventory.

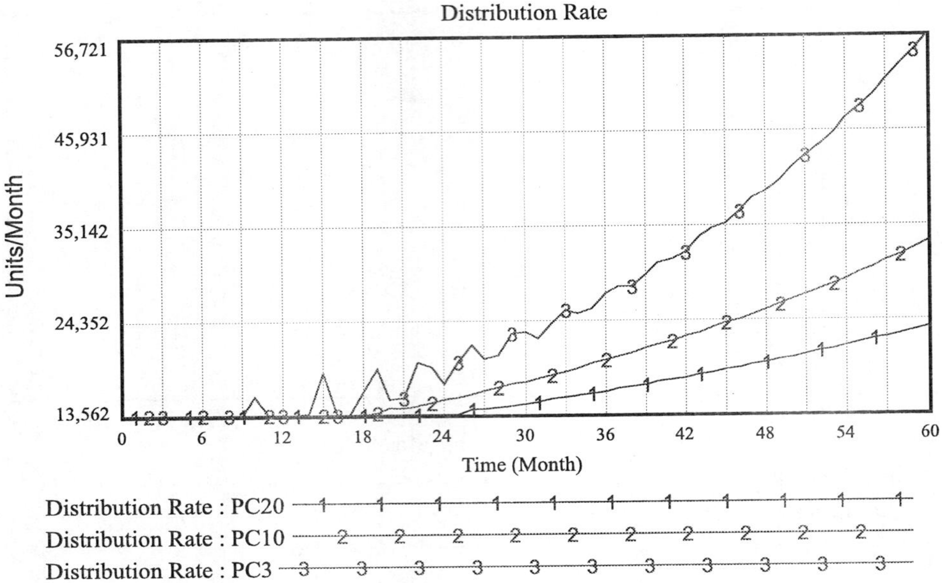

Figure 7.16: Effect of Production Cycle Time on Distribution Rate.

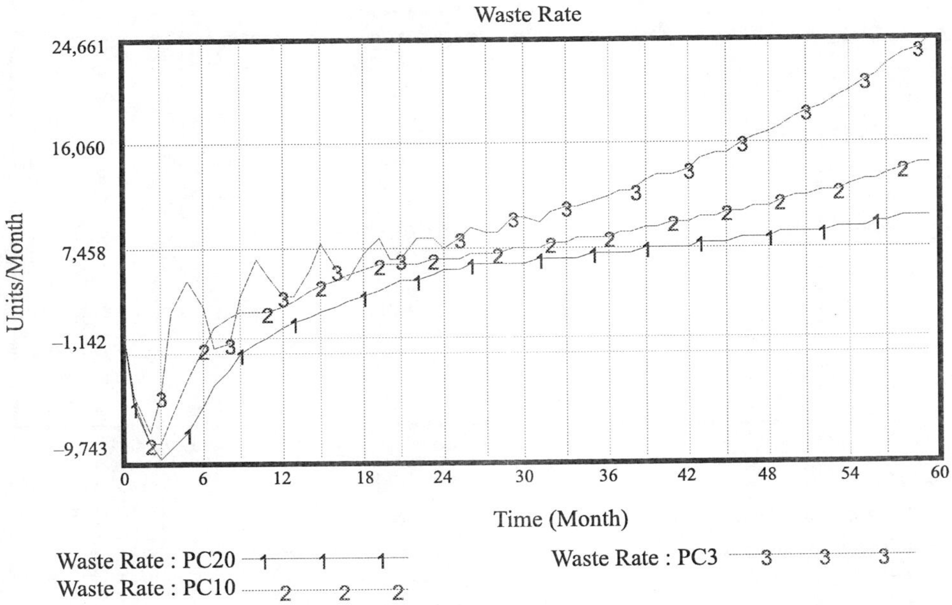

Figure 7.17: Effect of Production Cycle Time on Waste Rate.

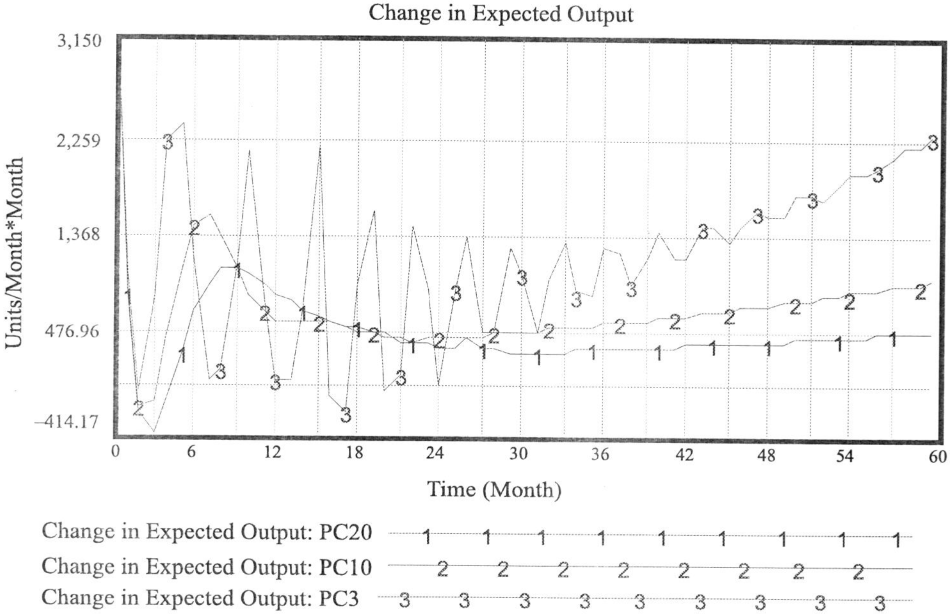

Figure 7.18: Effect of Production Cycle Time on Change in Expected Output.

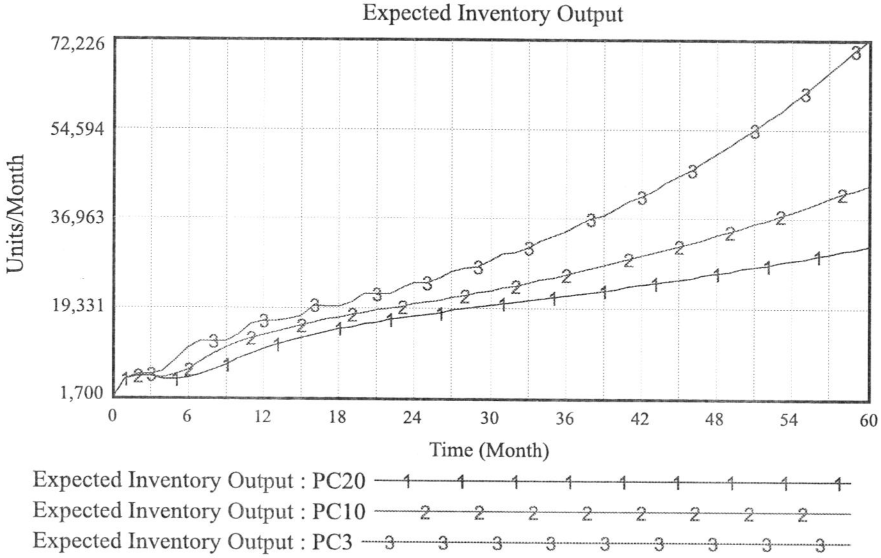

Figure 7.19: Effect of Production Cycle Time on Expected Inventory Output.

7.4 MODEL VALIDATION

Model validation is explained in detail in Chapter 9. However, this section provides some basic steps for model validation with specific reference to supply chain model. As this model has been developed from the earlier robust model of manufacturing SCM by Panov and Shiryaev (2003), the model has already passed the standard tests of validation. However, as the context of application in the present case is an utensil manufacturing industry, and also the addition of parameters can induce instability in the model structure, the simulation pattern has been superimposed with the empirical data obtained from the industry for a period of ten months for the sake of validation (Figure 7.20). It can be observed that the simulation follows the trend to a considerable extent even though there is deviation which could be due to the extraneous factors not considered in the model.

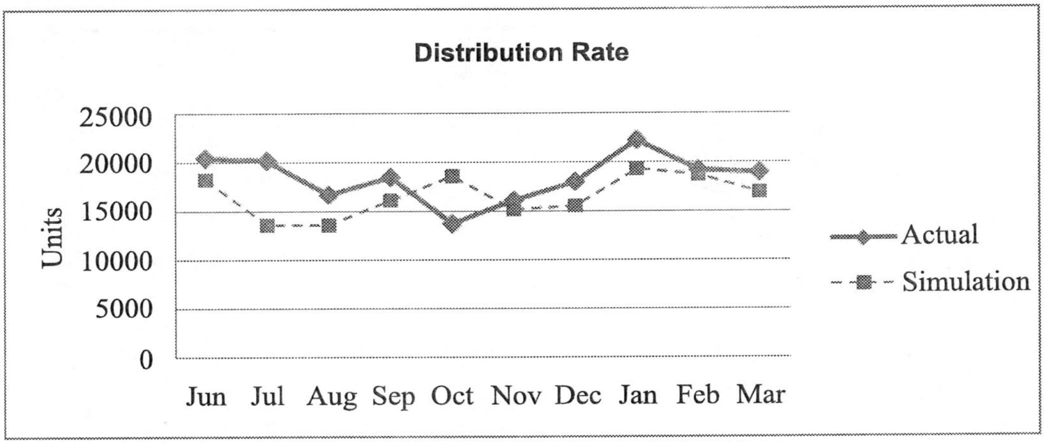

Figure 7.20: Actual Distribution Rate v/s Simulated Distribution Rate.

7.5 IMPLICATIONS OF THE STUDY

It is imperative from the study results that the firm should maintain a safety stock of at least 20,000 units to have a *stable production rate*, *work in progress*, *distribution inventory*, *distribution rate* and *waste rate* to stabilize the supply chain in 30 months.

Even though the study results are based on a small-scale aluminium industry, the model is generic in nature and can be used in medium- and large-scale industries, with minor changes applicable, as the key governing parameters remain the same.

7.6 CONCLUSION AND FUTURE WORK

This chapter considers the effect of various factors which influence the generic model of manufacturing supply chain and makes an attempt to minimize the 'bull-whip' effect. The policy implementation and recommendations of this research may be applied to save time and reduce cost involved in the supply chain management of any manufacturing industry. There is ample scope for further modification of the model by adding extraneous factors and factors which influence business cycle fluctuations, which have bearing on the supply chain efficiency.

7.7 SUMMARY

In this chapter, the development and deployment of a supply chain model has been undertaken. The model is developed using Vensim™ software and the readers may download this software from the Internet (freeware) and develop the model and run simulations and even modify some of the variables and study their influence on simulation results. The reader is advised to study the underlying theory of supply chain dynamics before developing a model. A thorough understandning of the Manufacture, Distributor and Retailer model would ease the process of model development. Also, an understanding of various delays involved in the supply and production would be necessary to undertake serious modelling exercise in SCM.

7.8 EXERCISE

Select a manufacturing plant of your choice say, automobile, machine tool, steel plant, televsion, garment, electronics, etc. and identify the supply chain variables. Using the basic model given in this chapter develop a causal diagram and stock and flow diagrams. Establish equations to relate various parameters chosen and run simulations to study the parameters of your interest. What would be your suggestion for enhancing the efficiency of SCM?

8

Innovation Dynamics in Manufacturing

Objectives

After going through this chapter you will be able to:

- Identify key variables, which influence Knowledge Management KM and innovation in manufacturing sector.
- Establish relationship between these variables.
- Develop a model considering the above two points and conduct simulation to study the influence of the dependent variables on performance of the manufacturing sectors.
- Implement policies to enhance the effectiveness of manufacturing.

8.1 INNOVATION DYNAMICS

In today's knowledge driven economy, Innovation is in the forefront of any Service or Manufacturing sector. Innovation is mainly a mind game, promoted through efficient Knowledge Management (KM), which is in turn a function of effectiveness of Human Resources in the organization. Innovation is a change in the thought process for doing something or new stuff that is made useful (McKeown, 2008). It may refer to an incremental, emergent or radical, and revolutionary changes in thinking, products, processes, or organizations. Innovation is an important topic in the study of economics, business, entrepreneurship, design, technology, sociology, and engineering. There are several variables which influence innovation and it is to some extent context based. So, the study of influencing factors of innovation through system dynamics can be very useful for policy-makers to have an idea about which are the factors which need close monitoring so as to enhance innovation.

It can be observed that innovation needs to be promoted at the Top Management level and well supported at the Plant level in a manufacturing sector. Figure 8.1 illustrates the variables which promote innovation at these two levels.

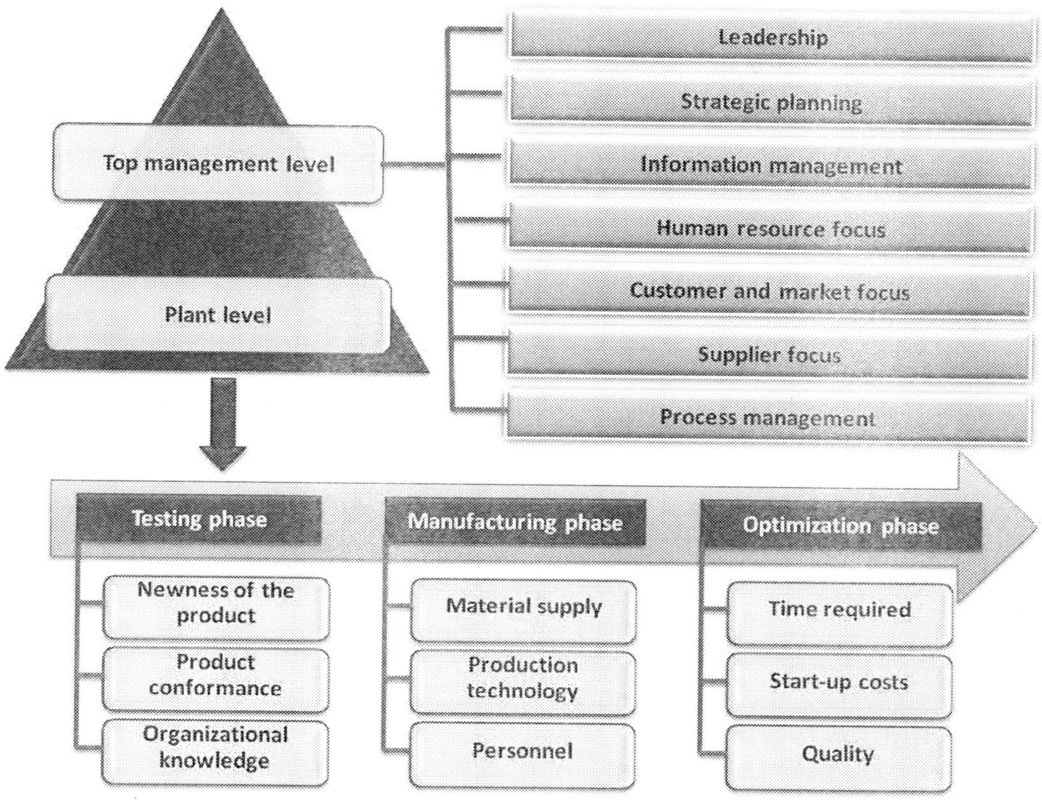

Figure 8.1: Key Variables that Affect KM and Innovation in Manufacturing Sector.

8.2 MODEL DEVELOPMENT

1. ***Problem Identification and Definition:*** The first phase of the SD methodology is the problem statement. The problem identified is the study of the influence of various factors on innovation so as to make recommendations on policy implementation.
2. ***The Dependence Diagram:*** The Dependence diagram indicates the causal relation between an independent variable and dependent variable. There can be more than one independent or dependent variables, and hence, sometimes it is even referred to as causal tree diagram. Before developing the model, it is very important to have a clear understanding of all the associated variables and their nature of interdependencies. Figure 8.2 depicts the Dependence diagram of variables. It can be observed that knowledge gain through SECI influences the total knowledge of the organization. Innovative ability, Rate of innovation success, and rate of knowledge transfer influences the capacity of innovation. Similarly, other interrelations have been shown in Figure 8.2.

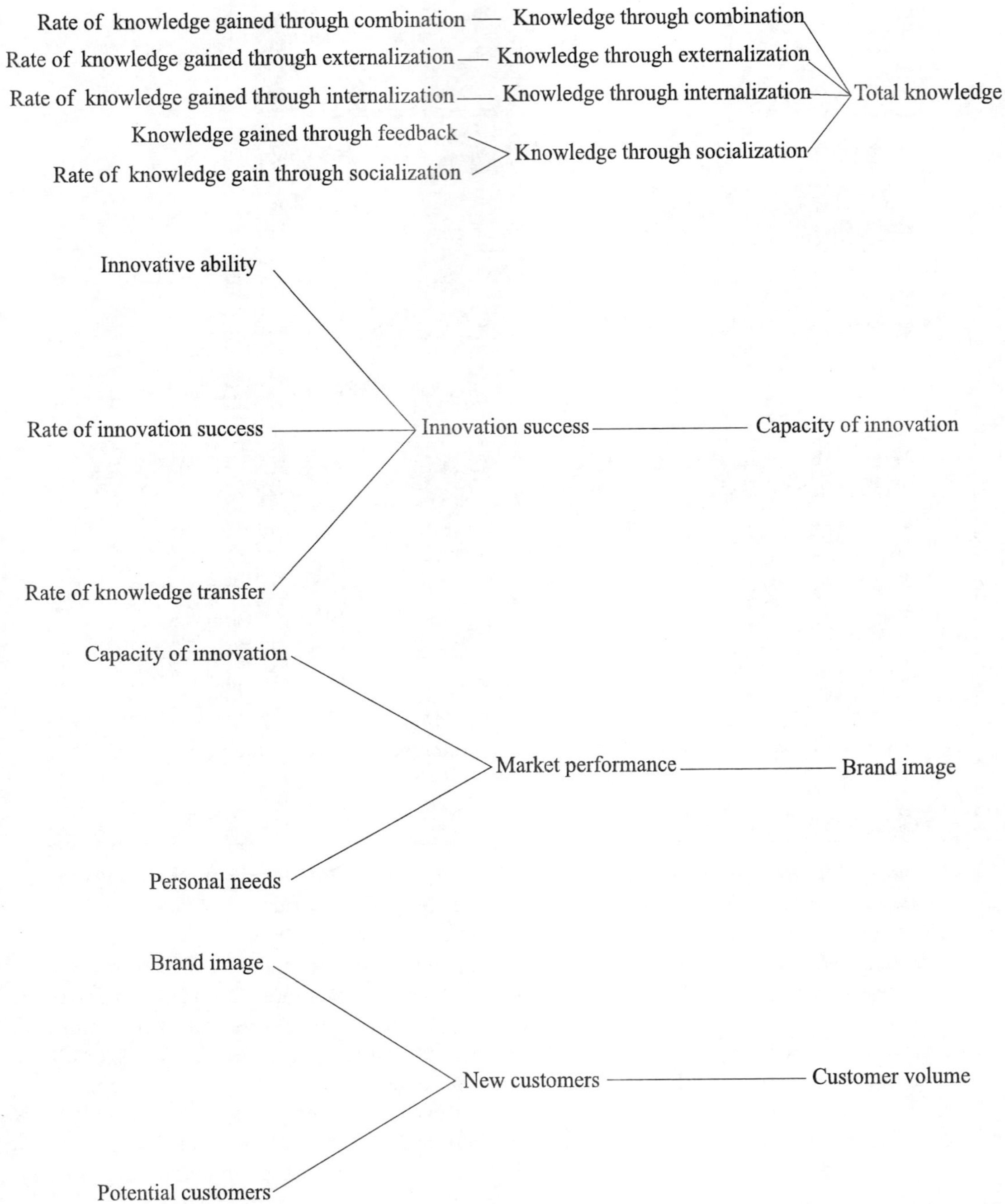

Figure 8.2: The Dependence Diagram of the Variables.

3. *System conceptualization:* The second phase of the methodology is the system conceptualization. After defining a problem a Causal Loop Diagram (CLD) is drawn (Figure 8.3). The diagram consists of a set of nodes representing the variables connected together. The relationships between these variables, represented by arrows, can be labelled as positive or negative. Three reinforcing loops can be observed in the CLD.
4. *Model formulation:* The third phase of the methodology involves model formulation. Unlike causal loop diagrams which characterize the initial view of the problem and basically serve the purpose of communication between the modeller and the policy-maker, the formulation of an operational model of the system is based on more specific structural details, like rates or policy variables, accumulation or level, auxiliaries, constants, information flows and delays. Flow diagrams represent such details and specific aspects of the model structure (Figure 8.4).

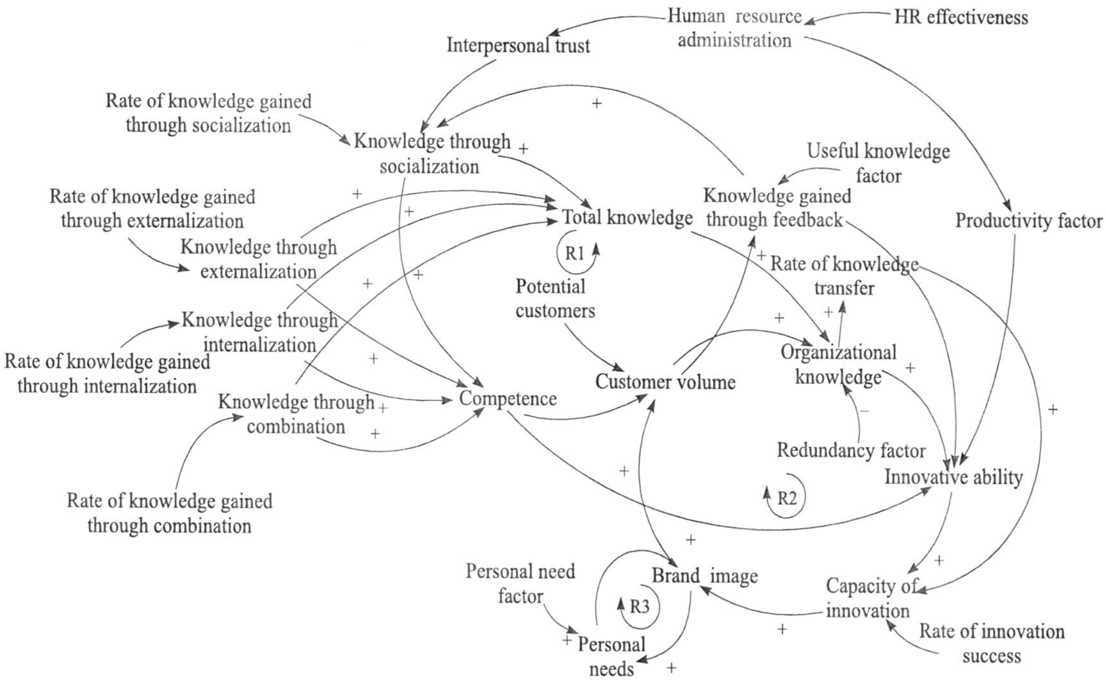

Figure 8.3: Causal Loop Diagram (CLD) of the Innovation Model.

Unlike causal loop diagrams which characterize the initial view of the problem and basically serve the purpose of communication between the modeller and the policy-maker, the formulation of an operational model of the system is based on more specific structural details, like rates or policy variables, accumulation or level, auxiliaries, constants, information flows and delays. Flow diagrams represent such details and specific aspects of the model structure (Figure 8.4).

5. *Simulation and validation:* Having identified the antecedents of innovation and KM in manufacturing, simulation can be performed to study the influence of the parameters on manufacturing effectiveness.

6. *Policy analysis and improvement:* Depending upon the simulation results, the factor playing a very significant role in the improvement of manufacturing effectiveness is considered and is fine tuned further to get the optimum level of system performance.

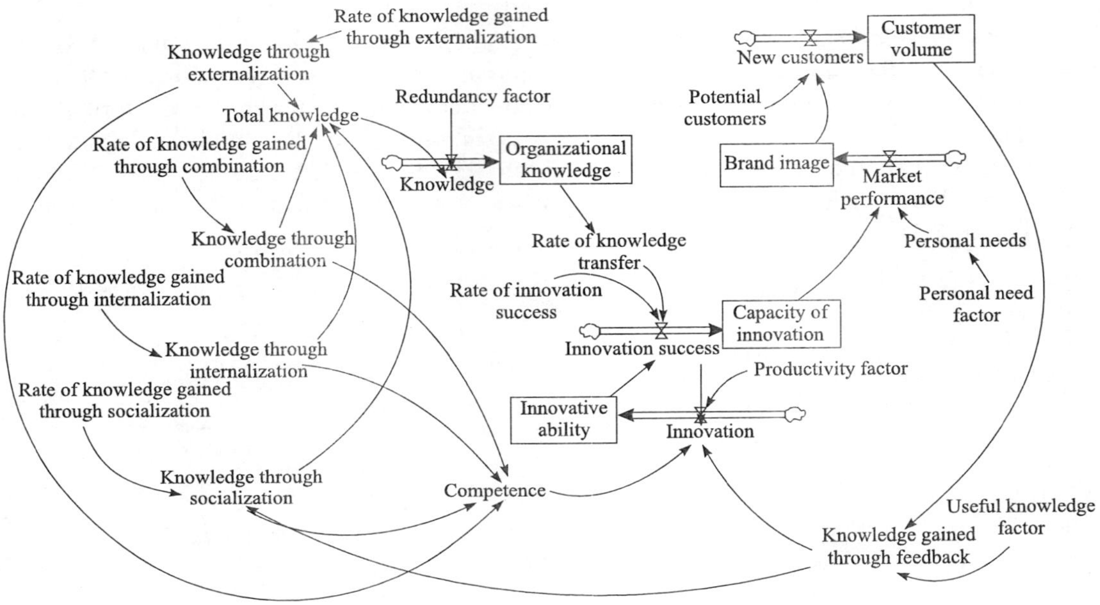

Figure 8.4: Stock and Flow Diagram of Innovation Model.

8.3 MODEL STRUCTURE

As briefly mentioned, a system dynamics approach to model the knowledge creation and acquisition dynamics within a manufacturing industry is considered.

8.3.1 Knowledge Conversion within a Firm

According to Nonaka and Takeuchi (Nonaka and Takeuchi, 1995), knowledge creation is the capability of a company as a whole to create knowledge, disseminate it throughout the organization, and embody it in products, services, and systems. They establish a dynamic model of knowledge creation. In this model, a critical assumption is that human knowledge is created and expanded through social interaction between tacit knowledge and explicit knowledge. This interaction is called 'knowledge conversion'. Knowledge conversion takes place through an iterative and spiral process of Socialization, Externalization, Combination and Internalization - or SECI - as an effective means of making individuals' tacit knowledge available to the broader organization in order to create new knowledge and then apply this new knowledge within their business processes towards achieving the organization's vision, objectives and performance standards (Figure 8.3).

1. In Socialization, the first stage, each person's tacit knowledge is converted to tacit knowledge now also held by other members in the firm. Socialization is primarily a process between individuals and occurs in settings such as apprenticeships and at conferences.
2. In the next stage, viz., Externalization, tacit knowledge is articulated and converted to explicit knowledge comprehensible to others, e.g. writing a report after attending a workshop. Externalization is a process among individuals within a group.
3. During Combination, the third stage, this newly developed explicit knowledge becomes widely disseminated, discussed, redesigned and modified. This is the area where information technology is most helpful, because explicit knowledge can be conveyed in documents, email, databases, as well as through meetings and briefings.
4. The final stage is Internalization. Internalization converts the changed, explicit knowledge again to a tacit form, this time held by many people. The internalization process is closely related to 'learning by doing' and transfers organization and group explicit knowledge to the individual.

8.3.2 Knowledge Transfer and Creation within an Industry

Similar to Nonaka's framework whereby tacit knowledge developed within one organizational unit is made explicit, transferred to another unit, applied within the new unit and thereby made tacit again; firms develop and transfer knowledge among themselves within their industry (Zack, 1999a). An organization develops tacit knowledge as a by-product of its activities. This knowledge may be made explicit to facilitate its transfer among other units of the organization. In doing so, it may leak out of the organization into the industry at large. At the same time, the organization may be absorbing knowledge leaking out of other firms within its industry, and internalizing that knowledge through its reapplication within the firm. It has to be noted that the knowledge created by the firm may not be entirely useful to the firm all the time, and hence its validation may be essential.

8.3.3 Competitive Knowledge and Capacity of Innovation

The loop for knowledge creation and transfer within an industry can be further extended with what Zack has defined as Strategic Knowledge—the knowledge the firm needs to execute its strategy (Zack, 1999b). He states that just a portion of the knowledge the firm possesses is strategically important as a source of sustainable competitive advantage, because such knowledge is highly valuable, unique to the organization, difficult to imitate and difficult to substitute (Barney, 1991). This introduces a factor called Redundancy factor. Redundancy factor is the factor which includes redundant data, incompletely processed data and other factors.

According to Zack (Zack, 1999b) every firm's strategic knowledge can be categorized by its ability to support a competitive position.

Core knowledge represents the basic knowledge required to operate in an industry and it is usually common to all members of an industry.

Advanced knowledge differentiates a firm, enabling it to compete and remain viable. Some firms compete head-on in particular knowledge domains, hoping that their knowledge is better than the competitors. Other firms seek to differentiate themselves based on what they know.

Innovative knowledge is truly unique and enables a firm to significantly differentiate itself from its competitors. It is obvious that knowledge is not static and what is innovative knowledge today will ultimately become the core knowledge of tomorrow. In the present model, the advanced knowledge and innovative knowledge are unified into a variable called *competence*. By employing this important resource, an organization is able to create a higher economic value for its customers. This in turn implies that the firm can gain more financial resources than its competitors (Almor and Hashai, 2004). The more financial resources the firm possesses, the more it can allocate to knowledge sharing and combination within the firm, and therefore, it can accelerate its learning cycle.

8.3.4 Affect of Innovation and Knowledge Management on Firm's Performance

The level of competencies is determined by the innovative capabilities of the people of the organization. Innovative capabilities grow from either deployment through learning-by-doing as achieved in innovation process, or by basic research. The logic in the model is that the innovative capabilities affect the ability of the firm to develop good products. Learning in turn affects the innovative capabilities. The learning function employs the logic of diminishing return so that the incremental learning from each decision is larger when the quantity of screening decisions is smaller.

The innovative capabilities, or the level of the knowledge stocks, affect the success rate of NPD and product launch. Adhering to the theory of absorptive capacity (Cohen & Levinthal, 1990), when the firm has above average technical capability, it produces better product ideas as the decision-makers are able to recognize technological opportunities better and develop novel solutions to fill customer need. It is described under the variable named as *productivity factor*.

8.4 SETTING UP SIMULATION

Using the causal loop diagram as a starting point, the model is set up for simulation in Ventana Systems VenSim® modelling environment. Despite the dynamic nature of system in general, the model has some constants, which reflect the assumptions made, to provide the basis for the model (Table 8.1). Using constants eases the modelling, but they also create an error source of their own. Assumptions behind constants attributes should be made logically so that the model stays intact. The constants presented in Table 8.1 realize that they are mostly market and industry related assumptions, even to that extent that these factors contain all the determinants of firm's business environment built in the model. This enables that the model can be customized relatively easily to new market settings by altering these constants. The numbers used in this simulation are based on rough estimates from experience from business cases, and aim to replicate a sort of general industrial firm. Thus, the results are also reported mainly for the purpose of highlighting the dynamics of the model more than anything else.

The success of any product or a service is measured by its market share. The market share can be measured by the number of customers using the product. The simulation is controlled via *Productivity factors* and the *Rate of knowledge gained* through various methods. The Total *knowledge* gained by the firm is tied to the Competence of the people working in the organization. A Knowledge organization, when provided with a good working culture results in the increase of productivity of the employees and leads to creative thinking. This creative thinking is the mother of all innovations.

A company that innovates continuously and delights the customer will have goodwill amongst customers and this will get it a brand name. Once the brand name is achieved, it becomes a household name, and hence, starts to attract new customers and increase its market share.

8.4.1 Exercising the Model

In this section, we describe how the system dynamics model is used to simulate different knowledge strategies that influence the knowledge level of the firm. To simulate the effect of adopting conservative or aggressive knowledge strategies, we change the various factors that activate the knowledge creation and the external knowledge acquisition loops respectively.

Four runs were simulated; the first run called as base run, which was a case of bad HR governance. The firm under consideration displayed poor Human Resource and Knowledge Management. The second, third and the fourth runs considered variation in the rates of knowledge gain through *Socialization, Redundancy factor, Rate of innovation success, Productivity factor* and *Useful knowledge factor*. A comprehensive study was done to learn the behaviour of the system by varying the factors.

Table 8.1: Variables and Constants used for Simulation

Variable Names		*Base run*	*Trial 1*	*Trial 2*	*Trial 3*
1.	Rate of knowledge through externalization	0.2	0.2	0.2	0.2
2.	Rate of knowledge through internalization	0.25	0.25	0.25	0.25
3.	Rate of knowledge through combination	0.25	0.25	0.25	0.25
4.	Rate of knowledge through socialization	0.1	0.3	0.4	0.5
5.	Redundancy factor	0.6	0.6	0.7	0.8
6.	Rate of innovation success	0.1	0.2	0.25	0.3
7.	Productivity factor	0.1	0.15	0.2	0.3
8.	Personal need factor	0.4	0.4	0.4	0.4
9.	Useful knowledge factor	0.1	0.25	0.3	0.35
10.	Potential customers	5000	5000	5000	5000

8.5 RESULTS

Simulation is performed to study the influence of the study parameters under four distinct cases.

1. Base run—Inferior HR&KM Strategy

Managing resourceful humans requires a constant balancing between meeting the human aspirations of the people and meeting the strategic and financial needs of the business. At times the balance can shift too far in one direction. In this case, a scenario of a manufacturing firm with no good practices of HR Management was considered. The parameters of *Rate of Knowledge gained through socialization*, *Rate of innovation success*, *Productivity factor* and *Useful knowledge factor* are kept minimum at 0.1. This indicates that the HR department is not considering the need of an effective knowledge management system and there is an initial knowledge gap. Moreover, the assumption is that the firm has a very bad employee morale and motivation. Other than this, it is assumed that all other factors were constant. The results are shown in figures below (Figures 8.5 to 8.9). This result suggests that the various factors under study take a very long time to attain the required level of performance.

2. Trial 1—Conservative HR&KM Strategy

The case considered in Trial 1 is that of a conservative HR&KM strategy. The manufacturing firm sets HR&KM policies and starts implementation of the same. Due to a sudden implementation of a new set of rules, there would certainly be a resistance to change. The workers mayn't embrace the new rules and the management decisions. Barriers to knowledge sharing have to be overcome which takes time. The parameters of Rate of knowledge gained through socialization, Rate of innovation success, Productivity factor and Useful knowledge factor are kept minimum at 0.3, 0.2, 0.15 and 0.25 respectively. The results are shown in figures below (Figures 8.5 to 8.9). This result suggests that, in this situation, the organizational knowledge starts to accumulate as soon as the employees are willing to work towards the KM objective. But, the outcomes of implementation of a good KM strategy, viz., Innovation capacity of the firm and its after effects on market performance will take time. This time delay will be due to streamlining of the system.

3. Trial 2—Moderate HR&KM Strategy

The case considered in Trial 2 is that of a moderate HR&KM strategy. The manufacturing firm sets HR&KM policies and the response it receives in due course is moderate. Over a period of time, slowly the attitude of people changes and they start to get themselves associated with the new strategies. The knowledge gaps disappear, socialization takes place amongst employees, exchange of ideas, experiences take place and hence the productivity gradually improves. This leads to a change in the organizational system and the work culture becomes healthy. The parameters of Rate of knowledge gained through socialization, Rate of innovation success, Productivity factor and Useful knowledge factor are kept at 0.4, 0.25, 0.2 and 0.3 respectively. The results are shown in figures below (Fig. 8.5 to 8.9). This result suggests that, in this situation the rate of accumulation of organizational knowledge is higher than that observed by implementing conservative HR&KM strategy. However, it was observed that, by adopting a moderate HR&KM strategy, the firm did much better than that of the conservative case.

4. Trial 3—Aggressive HR&KM Strategy

The case considered in Trial 3 is that of an aggressive HR&KM strategy. The manufacturing firm sets HR&KM policies and it is welcomed by all employees throughout the organization. This leads to a very high employee motivation, builds mutual respect towards each other and improves the employee morale. The HR department is believed to be really taking care of all the employee concerns and as a result all the employees of the organization feel a sense of ownership and work for the organization. Companies with good HR management will, over a period of time, leads to a healthy work culture. People in such organization discuss the problems encountered and get solutions from people who are experienced and hence effective knowledge sharing is ensured.

The parameters of Rate of knowledge gained through socialization, Rate of innovation success, Productivity factor and Useful knowledge factor are kept at 0.5, 0.3, 0.3 and 0.35 respectively. The results are shown in figures below (Figures 8.5 to 8.9). This result suggests that, an effective Human Resources and Knowledge Management will be the prime driver of innovation and creativity for any organization.

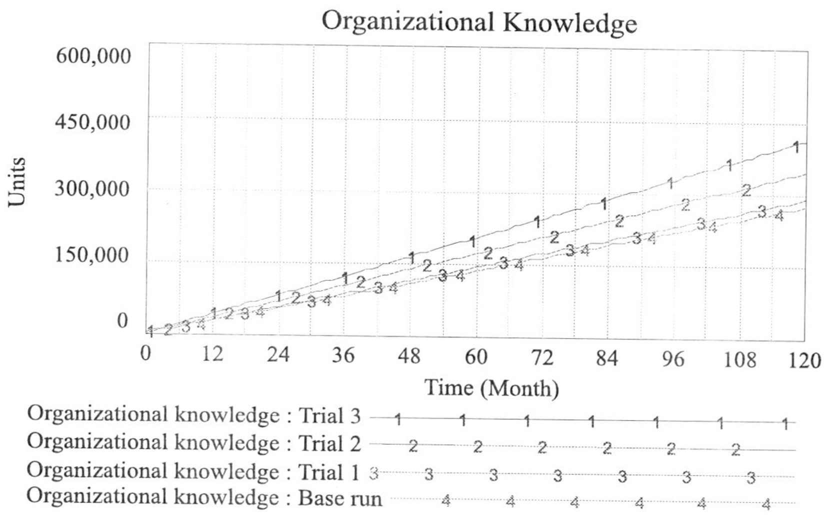

Figure 8.5: Effect of HR&KM Strategy on the Firm's Organizational Knowledge.

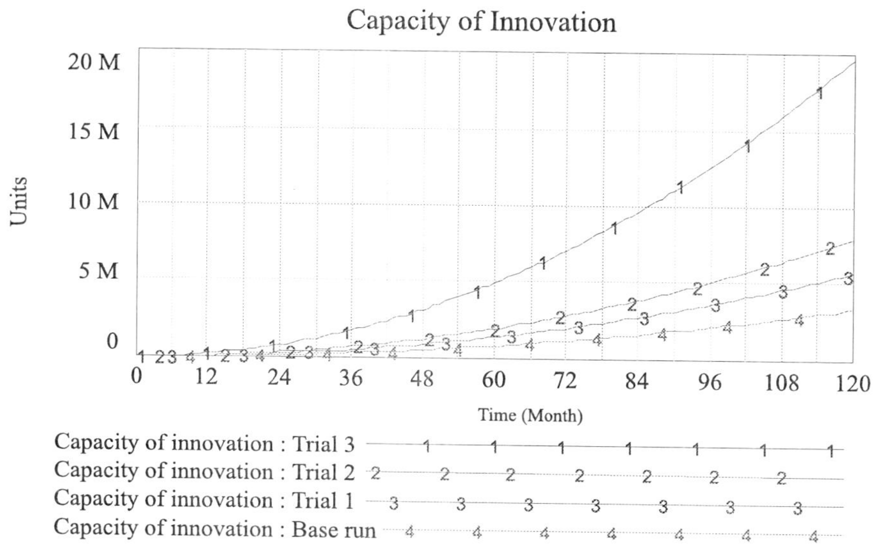

Figure 8.6: Effect of HR&KM Strategy on the Firm's Capacity of Innovation.

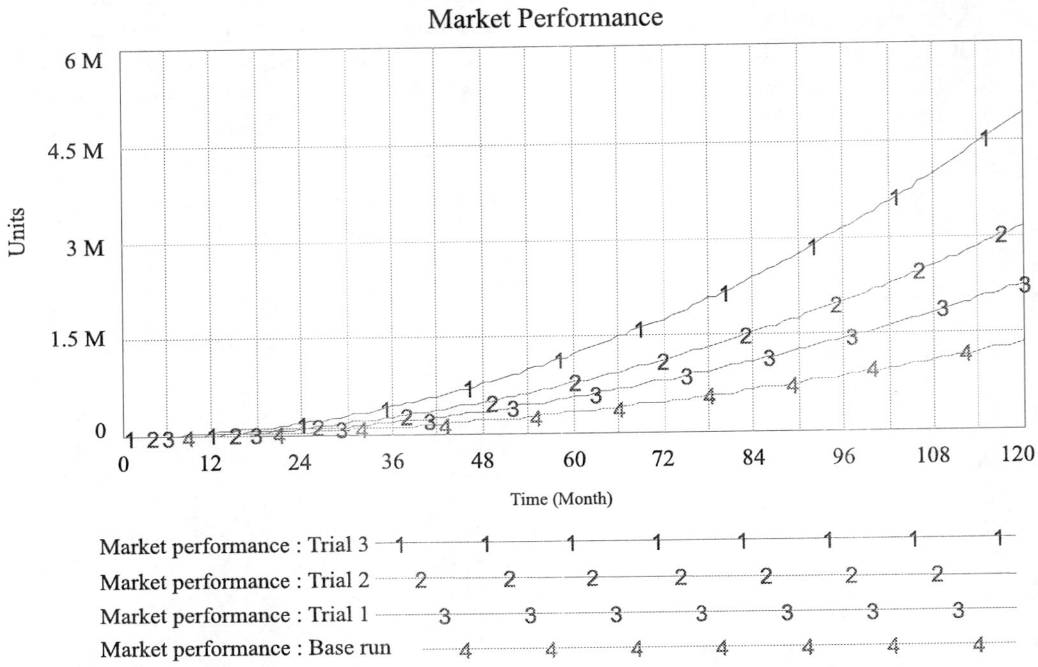

Figure 8.7: Effect of HR&KM Strategy on the Firm's Market Performance.

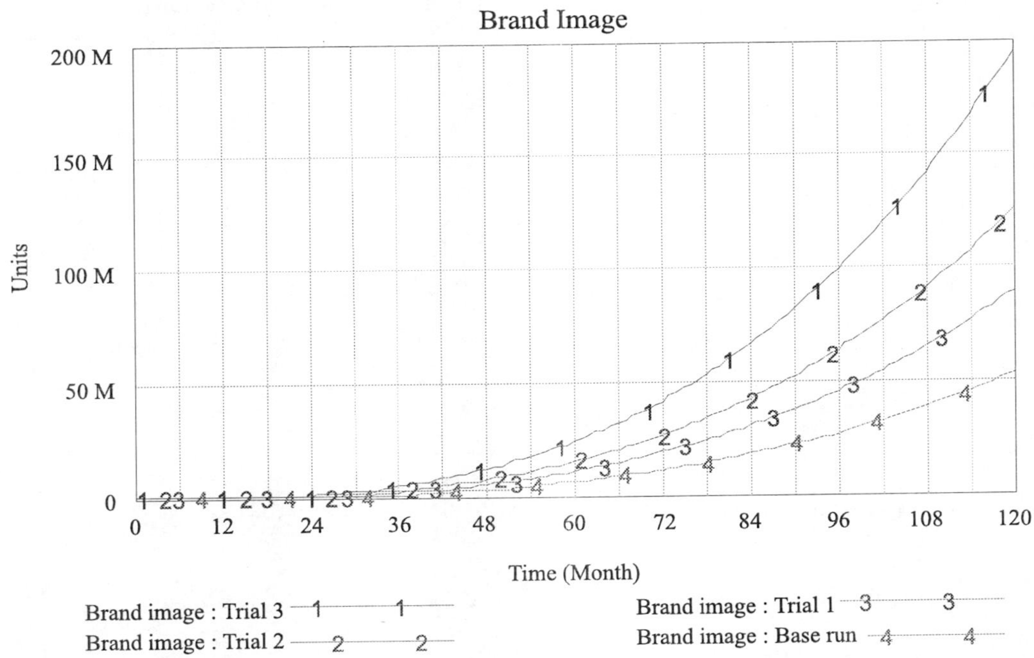

Figure 8.8: Effect of HR&KM Strategy on the Brand Image.

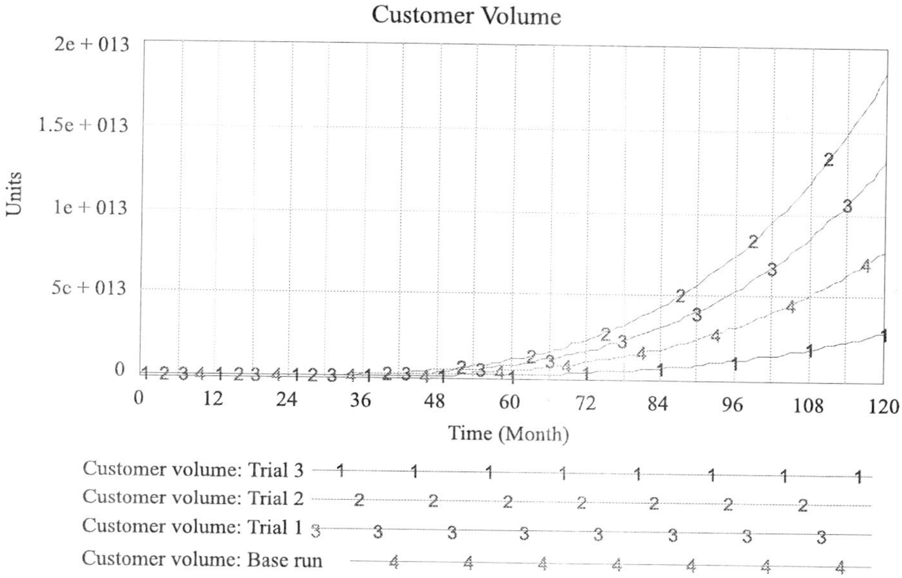

Figure 8.9: Effect of HR&KM Strategy on the Overall Customer Volume.

8.6 POLICY RECOMMENDATION

Globalization has shrunk the world in terms of accessibility but extended the boundaries of operation. It has brought about a new world order and is gradually reaching new heights, integrating all the fields to form a cohesive network. Its impact has touched political, cultural, economic, and ideological dimensions. Gone are the days when the consumers had to buy what was available in the market. The scenario has changed drastically over the years and now, the world has become a customer driven economy. This has resulted in more and more organizations locking themselves in fierce competitive combat to survive. For good or for ill, Darwin's 'survival of the fittest' theory is applicable even in the realm of business competition. This would hold well in the coming ages, and hence, to survive every organization has to consistently innovate.

The ability to innovate should be the vital core competency of any organization. Innovation requires extensive knowledge sharing and learning behaviours of an organization, leading to better performance in business process improvement and product and service offerings of a firm.

From the results it is evident that as in every case there are certain barriers which a HR department has to overcome whenever a change in the system is proposed. These are:

Parochial self-interest

- Individuals are more concerned with the implications for themselves

Misunderstanding

- Communication problems
- Inadequate information

Low tolerance of change

- Sense of insecurity
- Different assessment of the situation

Disagreement over the need for change

- Disagreement over the advantages and disadvantages.

A successful HR system is that which has the ability to convince and motivate the employees of the organization as to "Why the changes are made?" and "How it is going to help the organization and the people working in it?" Proper methods of rewards, recognition and incentives may motivate the employees in contributing significantly. Similarly, in the case of implementation of a KM strategy, it was seen that the case of aggressive HR and KM strategy was well ahead of the other strategies in terms of the key performance indicators of the firm's success like Market performance, Brand image, etc.

Even though, it is seen from the results that Organizational knowledge starts to improve on implementation of KM strategy, it takes time to streamline and bear fruits and reflect on the organizational performance. As observed in the results, it can be concluded that all the factors like Capacity of innovation, Market performance, Brand image and Customer volume have a very slow but a gradual growth in the initial years of implementation, but over a period of time the growth becomes exponential. Knowledge Management will be in the forefront of any organization as it is the prime driver of innovation and creativity and a good HR system is required to ensure the same.

8.7. SUMMARY

This chapter lists the key variables which influence KM and innovation in manufacturing sector, establishes the relationship between these variables and develops an SD model so as to study the influence of these variables on knowledge creation and acquisition by taking into consideration the Nonaka's model of knowledge transfer. Effect of Innovation and KM with specific reference to Inferior, Conservative, Moderate and Agressive HR and KM strategy on the manufacturing firm's performance has been studied through simulation results.

8.8. EXERCISE

Develop an SD model to study the influence of Technology Management on Manufacturing. The variables of KM studied in this chapter may be taken as the reference and from sources such as journals and international conference proceedings the variables of Technology management may be collected, their relationships may be established and an SD model may be developed to study the influence of each of these variables on manufacturing performance.

9

Validation of System Dynamics Model

Objectives

After going through this chapter you will be able to:

- Appreciate the necessity to conduct validation on a System Dynamics model.
- Identify various validation tests.
- Apply the method of conducting validation to a given problem situation.

Notation/Abbreviation

ECOPM = Engineering Competence Pool Model
RCRATE = Recruitment Competence Rate
TCRATE = Training Completion Rate
ALCA = Actual Level of Competence Absorbed
IOBPCS = Inventory and Order Based Production Control Structure
T_r = Competence Recovery Time
T_a = Forecast Competence Loss Rate Averaging Time
T_s = Skill Gap Recovery Time
T_{sl} = Skill Loss Rate Averaging Time
T_{td} = Delay in Training Rate
T_{dt} = Competence Training Delay Time

9.1. INTRODUCTION

Validations of Optimization Models, Decision Theory or Game Theory are often not questioned since the solution procedures are elegant and correct. But, in case of a System Dynamics model its effectiveness will simply be based on the credibility of the validation procedure. No model can be accepted unless it has passed the tests of validation. The validation procedures are context based and dynamic, but a systematic procedure must be followed by a modeller in order to authenticate the model. Reasons for system dynamic models having received more criticism could be due to the ease with which the models and their overall results are understandable. Another reason is that the simulation model of any system could only be an approximation of the actual system irrespective of the time spent on model building. Hence, if the model produced is not a 'close' enough approximation to this actual system, conclusions derived from such a model are likely to be divergent and erroneous; leading to possible costly decision mistakes (Ijeoma et al., 2001).

Validation is the process of establishing confidence in the usefulness of a model (Coyle, 1977). The process of validation can determine whether the model on which simulation is based, is an acceptably accurate representation of reality (Giannanasi et al., 2001). Validation can be done for all models regardless of whether their corresponding systems exist presently or would be built in future (Law & McComas, 2001). Also, Sterman (1984) gives insight on validation of models using statistical techniques and reasons that the technique that should be applied would depend on the availability of data in the real system. Law & Kelton (1991) contradicting the above authors, state that there is no such thing as an absolutely valid model. "Credibility of a model can be claimed only for the intended use of the model or simulation and for the prescribed conditions under which the model or simulation has been tested" (DMSO, 1996). Further, some authors like Sterman (2000) also argue that complete validation of a System Dynamics model is impossible; the emphasis should be more on model testing, i.e., the process to build confidence that a model is appropriate for the intended purpose.

There is no single test that would allow the modellers to assert that their models have been validated. Rather, the level of confidence in the model can increase gradually, as the model passes more tests (Forrester & Senge, 1980).

The modeller cannot carry out validation alone (Goldberg et al., 1990) and communication with the client or user plays an important role in building a valid model and establishing its credibility (Carson, 1989).

A wide range of tests to build confidence in a model have been developed by authors like Forrester and Senge (1980) and Barlas (1989, 1996). These tests prove the validity, credibility and generality (Solberg, 1992) of a model.

9.2. MODEL VALIDATION PROCESS

Schlesinger, (1979) has identified the key elements of the model validation process. The 'problem entity' is the system (real or proposed) to be modelled; the 'conceptual model' is the mathematical/verbal representation (flow diagram) of the problem entity developed for a particular study; and the 'computerized model' is the conceptual model implemented on a computer (control loop diagram). The inferences about the problem entity are obtained by conducting simulations on the computerized model in the experimentation phase.

The three steps in deciding if a simulation is an accurate representation of the actual systems considered are: verification, validation and credibility (Law & Kelton, 1991). Sargent (2003) defines 'Conceptual model validation' as the process of determining that the theories and assumptions underlying the conceptual model are correct and that the model representation of the problem entity is "reasonable" for the intended purpose. 'Computerized model verification' is the process of determining that a model implementation accurately represents the developer's conceptual description of the model and the solution to the model (AIAA, 1998). 'Operational validation' is the process of confirming that the model's output behaviour has sufficient accuracy for the model's intended purpose over the domain of the model's intended applicability. Operational validity determines the model's credibility. 'Data validity' ensures that the data necessary for model building, model evaluation and conducting the model experiments to solve the problem are adequate and correct (Love et al., 2000).

9.3. METHODS FOR TESTING AND VALIDATING THE SYSTEM DYNAMICS MODEL

The validation scheme as proposed by Forrester & Senge (1980) is enumerated below:

9.3.1. Importance of Model Objective

The validity of a model cannot be greater than the objective set for it. Therefore, the model objective must be a justified representation of the values prevalent in the real system. A model could be proven valid by a series of methods, but the validation may be totally useless if the objectives are wrongly set (Sushil, 1993).

9.3.2. Validating Model Structure

Tests of Suitability

- ***Structure-Verification Test:*** This test is meant to answer the following question: "Is the model structure not in contradiction to the knowledge about the structure of the real system, and have the most relevant structures of the real system being modelled?"
- ***Dimensional-Consistency Test:*** "Do the dimensions of the variables in every equation balance on each side of the equation?"
- ***Extreme-Conditions Test:*** "Does every equation in the model make sense even if subjected to extreme (but possible) values of variables?"
- ***Boundary-Adequacy Test:*** This test verifies whether the model structure is appropriate for the model purpose (Barlas, 1989).

Tests of Consistency

- ***Face Validity Test:*** "Does the model structure look like the real system? Does a reasonable fit exist between the feedback structure of the model and the essential characteristics of the real system?"
- ***Parameter-Verification Test:*** The numerical values of parameters should have real system equivalents.

Test of Utility and Effectiveness

- ***Appropriateness for the Audience:*** The more the appropriate a model for the audience, the more will be the audience's perception of model validity.

9.3.3. Validating Model Behaviour

Tests of Suitability

- ***Parameter Sensitivity Test:*** "Do the modes of behaviour change with the parameter variations?"

- ***Structural Sensitivity Test:*** "Is the behaviour of the model sensitive to reasonable structural reformulation?"

Tests of Consistency

- ***Behaviour-Reproduction Test:*** Here, the generated model behaviour is judged with the historical behaviour pattern.
- ***Behaviour-Prediction Test:*** "Whether or not the model generates patterns of future behaviour in terms of periods, shape or other characteristics?
- ***Behaviour-Anomaly Test:*** "What behaviour shown by the model is conflicting with the real system behaviour and how implausible behaviour arises if the assumptions are altered?"
- ***Family Member Test:*** Parameter values are chosen to depict a particular situation. By choosing a different set of parameter values can the model be applied to other situation as well?
- ***Surprising Behaviour Test:*** "Does the model under some test circumstances produce dramatically unexpected or surprise behaviour, not observed in the real system?
- ***Extreme-Policy Test:*** If the model behaves in an expected fashion even under extreme policies, then it boosts confidence in the model (Saysel et al., 2004).
- ***Boundary Adequacy Test:*** If the extra model structure does not change the behaviour, then this extra structure is not necessary. Alternatively, if a model structure does not reproduce desired model behaviour, it calls for inclusion of additional model structure (Barlaś, 1996).
- ***Behaviour-Sensitivity Test:*** "Does plausible shift in parameters cause model to fail behaviour tests previously passed?" (Saysel et al., 2004).
- ***Statistical Tests:*** "Does the model pass statistical tests based on the data from real system?"

Test of Utility and Effectiveness

- ***Counter-Intuitive Behaviour:*** "In response to some policies, does the model exhibit behaviour that at first contradicts intuitions, and later, with the aid of the model, is seen as a clear implication of the structure of the system?" (Richardson et al., 1981).

9.3.4. Validating Policy Implications

Test of Suitability

- ***Policy Sensitivity and Robustness Test:*** "Does the model based policy recommendations change with reasonable changes in the parameter values, or reasonable alteration in the equations?"

Tests of Consistency

- ***Changed Behaviour Prediction Test:*** "Does the model correctly predict how the behaviour of the system would change if the governing policy is changed?"
- ***Boundary Adequacy Test:*** "Does modifying of the model boundary alter policy recommendations?"

- ***System Improvement Test:*** "Are the policies found beneficial after working with a model?

Test of Utility and Effectiveness

- ***Implementable Policy Test:*** "Can those responsible for policy-making in the real system be convinced of the values of model-based policy recommendations?

9.4. A CASE STUDY

Consider the Engineering Competence Model (ECOPM) developed in the Chapter 5. The systematic validation procedure may be carried out as follows:

9.4.1. Importance of Model Objective

The objective of the ECOPM is to analyze the dynamic response of performance indices namely: Recruitment Completion Rate (RCRATE), Training Completion Rate (TCRATE), and Actual Level of Competence Absorbed (ALCA) against the six time-based variables: Competence Recovery Time (T_r), Competence Training Delay Time (T_{dt}), Forecast Competence Loss Rate Averaging Time (T_a), Skill Gap Recovery Time (T_s), Delay in Training Rate (T_{td}) and Skill Loss Rate Averaging Time (T_{sl}).

9.4.2. Validating the Model Structure

Tests of Suitability

- ***Structure-Verification Test:*** The ECOPM considers only primary internal structures relevant to the model with no interference or input from the external world. Further, the structure of the model in this research considers all the real-life issues, which have significant influence on the competence pool, viz., RCRATE, TCRATE and ALCA.
- ***Dimensional-Consistency Test:*** The dimensions of all the variables were checked and the equations were verified and balanced, e.g.,

$$RCRATE = RATE* \frac{1}{1 \quad T_{rd} * S}$$

L.H.S. = R.H.S. = Units/month.

- ***Extreme-Conditions Test:*** The model has to be tested against extreme values, e.g., Figure 5.7 T_r is varied from 0 to 30 months. Beyond 30 months there is no significant change of behaviour.
- ***Boundary-Adequacy Test:*** As the model indicates, it considers all variables related to the system except for some independent variables, which are external to the system and whose absence will not create any difference in the final output for the defined boundary condition. Furthermore, ECOPM is a derivative of the IOBPCS model (Towill, 1982), which has been successfully validated by various authors.

Tests of Consistency

- ***Face Validity Test:*** The IOBPCS model was developed by Coyle (1977) to represent industrial practice in the UK. The ECOPM has been derived from the IOBPCS model with changes incorporating certain deciding variables but mainly involves human factors as in case of IOBPCS model.
- ***Parameter-Verification Test:*** The parameters correspond conceptually and numerically to real life. A maximum delay of 6 months for the retraining is based on the prevalent conditions in engineering education, and hence, stands valid.

Test of the Utility and Effectiveness

- ***Appropriateness for Audience:*** The model is easy to understand with simple feed-forward and feedback structures. All the terms used are appropriate to the context and easily understandable by the practitioners.

9.4.3. Validating the Model Behaviour

Tests of Suitability

- ***Parameter Sensitivity Test:*** As shown in the graph, the model has shown sensitivity to variations in time-based policy parameters (Figure 5.7).

  $$T_a = T_{dt} = 6 \text{ months}$$

  T_r is varied from 0 to 30 months.
- ***Structural Sensitivity Test:*** Any reformulation to the models structure would cause the results to change.

Tests of Consistency

- ***Behaviour-Reproduction Test:*** The patterns generated in the simulations are in tune with the patterns in real-life situations, such as: skill recovery rate, peak over shoot, etc.
- ***Behaviour-Prediction Test:*** This model can be successfully used for forecasting the competence pool pattern, e.g., the model shows the time required to recover the competence gap.
- ***Behaviour-Anomaly Test:*** No erratic behaviour was observed during the course of the simulation and hence, anomaly of any kind does not exist.
- ***Family Member Test:*** The ECOPM is derived from the IOBPCS model that could be included under the same family of models. Also, by choosing a different set of parameter values, it may be used for some other purpose.
- ***Surprising Behaviour Test:*** There was a case of peak overshoot observed (Figure 5.22). The cause for this phenomenon may be attributed to the delay caused within the system, which amounts to accumulation. So a surprising behaviour has a plausible explanation.
- ***Extreme-Policy Test:*** The extreme condition test is carried out for extremely high and low values of policy parameters and behaviour of the system observed. As it is observed from the graph (e.g., Figure 5.22), for extreme values of policy parameters, a dip in the ALCA is observed along with a delay in time for rise. Hence, it can be a good representation of real system.

- ***Boundary Adequacy Test:*** At the boundary values, the system behaviour has remained consistent as may be observed in the 3D-surface graphs in Chapter 5. So the predefined boundaries are adequate to explain the system behaviour.
- ***Behaviour-Sensitivity Test:*** The model is quite sensitive to the variations in the policy parameter as observed in the graphs.

Test of Utility and Effectiveness

- ***Counter Intuitive Behaviour:*** This test proves that the model does not exhibit any contradictive behaviour in response to some policies.

9.4.4. Validating Policy Implications

Test of Suitability

- ***Policy Sensitivity and Robustness Test:*** The results of this test are positive and the model is robust.

Tests of Consistency

- ***Changed Behaviour Prediction Test:*** This test is to check how behaviour of the system will change if a governing policy is changed.
- ***Boundary Adequacy Test:*** This test has shown that modifying the model boundary would alter policy recommendations.

Test of Utility and Effectiveness

- ***Implementable Policy Test:*** The critical policy parameters identified through simulation are logically justifiable. For example, the competence recovery period of 6 months is justifiable based on actual conditions.

9.5. SUMMARY

Validation of SDM is a complex process, but is unavoidable as it is the proof for the reliability and authenticity of the model. The process is complex and tiresome because no single procedure can fit into all the System Dynamics models. The discussion on validation procedure given in this chapter, as given by a group of researchers, provides insight into this field.

This chapter gives a standard procedure for testing and validating a System Dynamics model. The process is explained in detail through the ECOPM model developed in Chapter 5. The reader is encouraged to pass his/her model through as many tests as possible so as to make the simulation results acceptable.

9.6. EXERCISE

Conduct Model Validation for the problems in the previous chapters.

10

Mathematical Modelling and Simulation of Robot Arm

Objectives

After going through this chapter you will be able to:

- Understand the meaning of 'Mathematical Modelling', 'Singularity', and 'Kinematic Decoupling'.
- Understand and appreciate the importance of Mathematical Modelling and Simulation of any robot prior to investigations with the actual robot.
- Identify the D-H (Denavit-Hartenberg) parameters for the given robot, and hence, construct the Kinematic matrix matrices by homogeneous transformation of the D-H parameters, for the Forward Kinematics problem.
- Understand the scope of applicability of 'Kinematic Decoupling' employed here to solve the Inverse Kinematics problem of the Stanford Scheinmann arm.
- Digest the algorithm used to simulate the end effector movement of the robot arm and construct such similar algorithms for any given robot.

10.1 DEFINITIONS

Mathematical Modelling

The mathematical description of any given situation.

Singularity

Configuration for which certain directions of motion may not be possible (Craig and John, 1986).

Kinematic Decoupling

Treating the position and orientation of the end effector or the tool frame independently and being able to solve for all joint parameters independently from Inverse Kinematics. Decoupling is applicable for robot arms with a spherical wrist (Spong, et al., 2004).

10.2 INTRODUCTION

One of the advantages of mathematical modelling/simulation is that it gives an insight into the predicted motion or trajectory following of the robot in real time and also information regarding singularities of the robot in its workspace.

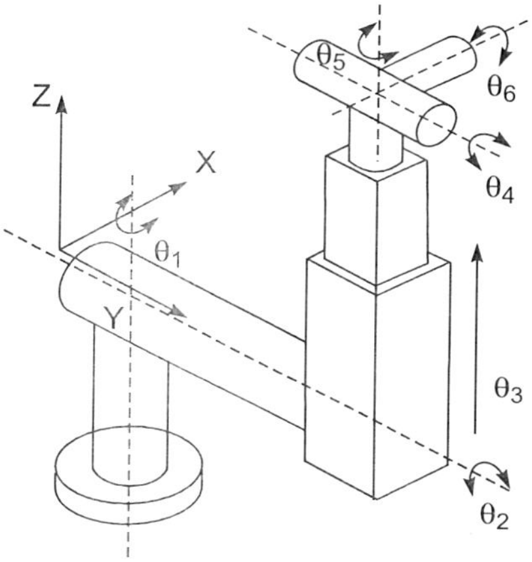

Fig 10.1: Stanford Scheinmann Arm.

Robots aid people in everyday life and by the next decade or so, they may even take over the daily routine proceedings of humans. Robots are currently very widely used – from toys through workplaces, industries and to very sophisticated ones needed for space exploration. The Stanford Scheinmann Arm was designed in 1969 at Stanford University. It is a 6 d.o.f. robot and all drives are electric. It finds its application in various fields like teaching, research or a suitable end effector can be attached for a desirous task to be performed. It is completely controlled by a computer. Brakes are used on all joints to hold the arm in position while the computer computes the next trajectory. Drives are DC electric motors, harmonic drives and spur gears. Electromechanical brakes are used for locking joints. It is very useful to perform simulation prior to investigations with the actual robot because simulations are easy to set up, less expensive, faster and more convenient to use. It allows a better design exploration. Final design can be verified before one thinks of investing in the manufacturing of the actual robot or even the prototype, for that matter. The Stanford Scheinmann Arm helped to develop the knowledge base which has been applied in essentially all industrial robots of today.

10.3 THE PROBLEM STATEMENT

Mathematical modelling of the Stanford Schienmann robot arm and to simulate its end effector movement by constructing an algorithm using MATLAB.

10.4 FORWARD KINEMATICS

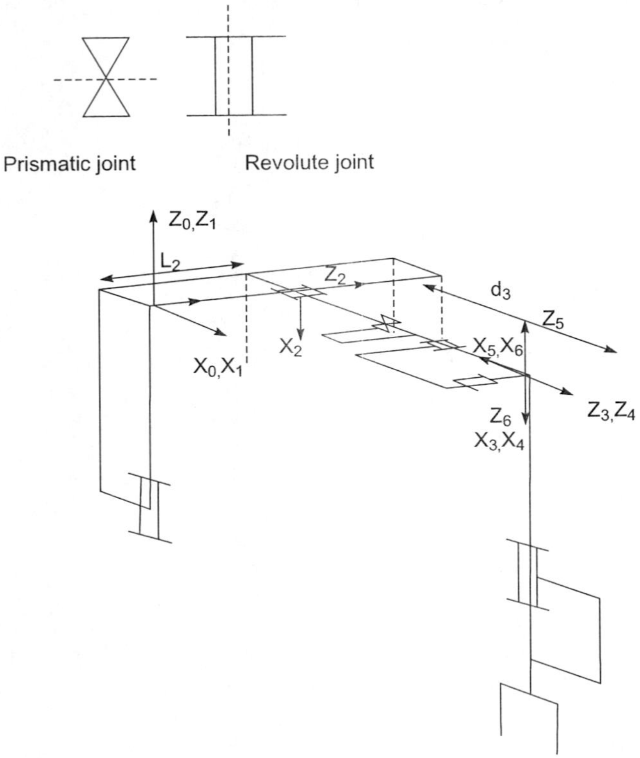

Fig 10.2: Schematic Diagram with Joint Axes and Link Length Descriptions.

D-H parameters are:

a_i = distance between the axes Z_i and Z_{i+1} along X_i

α_i = angle between the axes Z_i and Z_{i+1} about X_i

d_i = distance between the axes X_{i-1} and X_i along Z_i

θ_i = angle between the axes X_{i-1} and X_i about Z_i

X axis is chosen such that it is a common normal to the two successive frames

Table 10.1: D-H Parameters Configuration Table

Link no (i)	α_{i-1}	a_{i-1}	d_i	θ_i
1	0	0	0	θ_1
2	– 90	0	L_2	θ_2
3	90	0	d_3	θ_3
4	0	0	0	θ_4

Contd.

Contd.

5	90	0	0	θ_5
6	90	0	0	θ_6

Transformation /A – matrices:
[Here, c_1= cos(θ) and s_1= sin(θ)]

$$ {}^0_1T = \begin{bmatrix} c_1 & -s_1 & 0 & 0 \\ s_1 & c_1 & 0 & 0 \\ 0 & 0 & 1 & 0 \\ 0 & 0 & 0 & 1 \end{bmatrix} = A_1\ (\theta_1) $$

$$ {}^1_2T = \begin{bmatrix} c_2 & -s_2 & 0 & 0 \\ 0 & 0 & 1 & L_2 \\ -s_2 & -c_2 & 0 & 0 \\ 0 & 0 & 0 & 1 \end{bmatrix} = A_2\ (\theta_2) $$

$$ {}^2_3T = \begin{bmatrix} 1 & 0 & 0 & 0 \\ 0 & 0 & -1 & -d_3 \\ 0 & 1 & 0 & 0 \\ 0 & 0 & 0 & 1 \end{bmatrix} = A_3\ (\theta_3) $$

$$ {}^3_4T = \begin{bmatrix} c_4 & -s_4 & 0 & 0 \\ s_4 & c_4 & 0 & 0 \\ 0 & 0 & 1 & 0 \\ 0 & 0 & 0 & 1 \end{bmatrix} = A_4\ (\theta_4) $$

$$ {}^4_5T = \begin{bmatrix} c_5 & -s_5 & 0 & 0 \\ 0 & 0 & 1 & 0 \\ -s_5 & -c_5 & 0 & 0 \\ 0 & 0 & 0 & 1 \end{bmatrix} = A_5\ (\theta_5) $$

$$ {}^5_6T = \begin{bmatrix} c_6 & -s_6 & 0 & 0 \\ 0 & 0 & -1 & 0 \\ s_6 & c_6 & 0 & 0 \\ 0 & 0 & 0 & 1 \end{bmatrix} = A_6\ (\theta_6) $$

The kinematic matrix:

$$\mathrm{K} = {}^{0}_{N}T = {}^{2}_{1}T\ {}^{1}_{2}T \ldots {}^{N-1}_{N}T = A_1\,(\theta_1)\,A_2\,(\theta_2)\,\ldots\,A_N\,(\theta_N)$$

Thus,

$$\begin{aligned} {}^{0}_{6}T &= {}^{1}_{0}T\ {}^{1}_{2}T\ {}^{2}_{3}T\ {}^{3}_{4}T\ {}^{4}_{5}T\ {}^{5}_{6}T \\ &= A_1\,(\theta_1)\,A_2\,(\theta_2)\,A_3\,(\theta_3)\,A_4\,(\theta_4)\,A_5\,(\theta_5)\,A_6\,(\theta_6) \end{aligned}$$

Therefore,

$$\begin{aligned} \mathrm{K} &= {}^{0}_{6}T = {}^{0}_{1}T\ {}^{1}_{2}T\ {}^{2}_{3}T\ {}^{3}_{4}T\ {}^{4}_{5}T\ {}^{5}_{6}T \\ &= A_1\,(\theta_1)\,A_2\,(\theta_2)\,A_3\,(\theta_3)\,A_4\,(\theta_4)\,A_5\,(\theta_5)\,A_6\,(\theta_6) \end{aligned}$$

$$\begin{bmatrix} c_6[c_5(c_1c_2c_4 - s_1s_4) - s_5c_1s_2] - s_6(c_1c_2s_4 + s_1c_4) & -s_6[c_5(c_1c_2c_4 - s_1s_4) - s_5c_1s_2] - c_6(c_1c_2s_4 + s_1c_4) & s_5(c_1c_2c_4 - s_1s_4) + c_5c_1s_2 & c_1s_2d_3 - s_1L_2 \\ c_6[c_5(s_1c_2s_4 + c_1s_4) - s_5s_1s_2] + s_6(c_1c_4 - s_1c_2s_4) & -s_6[c_5(s_1c_2s_4 + c_1s_4) - s_5s_1s_2] + c_6(c_1c_4 - s_1c_2s_4) & s_5(s_1c_2s_4 + c_1s_4) + c_5s_1s_2 & s_1s_2d_3 + c_1L_2 \\ c_6(c_2s_5 - s_2c_4c_5) + s_6s_1s_4 & s_6(c_2s_5 - s_2c_4c_5) + c_6s_2s_4 & -(s_2c_4s_5 + c_2c_5) & -c_2d_3 \\ 0 & 0 & 0 & 1 \end{bmatrix}$$

The first three columns of the final kinematic matrix represent the rotation matrix of the tool frame or the end effector frame of the robot arm.

The last column of the matrix describes the end effector position of the robot arm with respect to frame zero.

10.5 INVERSE KINEMATICS

Since the direct approach of solving the inverse kinematics will be cumbersome, a method called kinematic decoupling is used. It is particularly helpful in this case because the last three links of the arm have a common centre (spherical wrist). So θ_1, $\theta_2 \ldots \theta_6$ are found given the final position.

For the first three joints:

$$\text{We have } {}^{0}_{3}T = \begin{bmatrix} c_1c_2 & -s_1 & c_1s_2 & c_1s_2d_3 - s_1d_2 \\ s_1c_2 & c_1 & s_1s_2 & s_1s_2d_3 + c_1d_2 \\ -s_2 & 0 & c_2 & c_2d_3 \\ 0 & 0 & 0 & 1 \end{bmatrix}$$

$$\text{Let us assume the given position of the end effector to be } = \begin{bmatrix} n_x & s_x & a_x & p_x \\ n_y & s_y & a_y & p_y \\ n_z & s_z & a_z & p_z \\ 0 & 0 & 0 & 1 \end{bmatrix}$$

Position here, means both the position and orientation of the end effector with respect to frame zero.

$$ {}^0_3T = {}^0_1T\ {}^1_2T\ {}^2_3T $$

$$ ({}^0_1T)^{-1}\ {}^0_3T = {}^1_2T\ {}^2_3T $$

Note: If $A = \begin{bmatrix} R & t \\ 000 & 1 \end{bmatrix}$ where R = rotation matrix, t = translation/position column

Then $$ A^{-1} = \begin{bmatrix} R^T & -R^T t \\ 000 & 1 \end{bmatrix} $$

Thus, $$ ({}^0_1T)^{-1} = \begin{bmatrix} c_1 & s_1 & 0 & 0 \\ -s_1 & c_1 & 0 & 0 \\ 0 & 0 & 1 & 0 \\ 0 & 0 & 0 & 1 \end{bmatrix} $$

$$ ({}^0_1T)^{-1}\ {}^0_3T = \begin{bmatrix} x & x & x & c_1p_x + s_1p_y \\ x & x & x & -s_1p_x + c_1p_y \\ x & x & x & p_z \\ 0 & 0 & 0 & 1 \end{bmatrix} \quad \ldots[1] $$

$$ {}^1_2T\ {}^2_3T = \begin{bmatrix} x & x & x & s_2d_3 \\ x & x & x & L_2 \\ x & x & x & c_2d_3 \\ 0 & 0 & 0 & 1 \end{bmatrix} \quad \ldots[2] $$

Equating (1) and (2),

$$ -S_1P_x + C_1P_y = L_2 \quad \ldots[3] $$

let $$ x = \sqrt{px^2 + py^2},\ \phi = \tan^{-1}\left(\frac{p_y}{p_x}\right) $$

i.e., $$ p_x = r\cos\phi,\ p_y = r\sin\phi $$

Substituting in (3),

$$ -s_1\,(r\cos\phi) + c_1\,(r\sin\phi) = L_2 $$

$$ r\sin(\phi - \theta_1) = L_2 $$

$$ \sin(\phi - \theta_1) = \frac{L_2}{r} $$

Therefore, $$\cos(\phi - \theta_1) = \pm\sqrt{1-\sin^2(\phi-\theta_1)}$$

$$= \pm\sqrt{\frac{r^2 - L_2^2}{r}}$$

Therefore, $$\tan(\phi - \theta_1) = \pm\frac{L_2}{\sqrt{r^2 - L_2^2}}$$

$$\phi - \theta_1 = \tan^{-1}\left(\frac{\pm L_2}{\sqrt{r^2 - L_2^2}}\right)$$

Therefore, $$\theta_1 = \phi - \tan^{-1}\left(\frac{\pm L_2}{\sqrt{r^2 - L_2^2}}\right)$$

$$\theta_1 = \tan^{-1}\left(\frac{p_y}{p_x}\right) - \tan^{-1}\left(\frac{\pm L_2}{\sqrt{r^2 - L_2^2}}\right) \quad \ldots[4]$$

Similarly, $S_2 d_3 = c_1 p_\text{x} + s_1 p_\text{y}$ and $c_2 d_3 = p_z$

Therefore, $$\frac{s_2}{c_2} = \tan\theta_2 = \frac{c_1 p_x + s_1 p_y}{p_z}$$

Or $$\theta_2 = \tan^{-1}\left(\frac{c_1 p_x + s_1 p_y}{p_z}\right) \quad \ldots[5]$$

As noted, $$s_2 d_c = c_1 p_x + s_1 p_y \quad \ldots[6]$$
$$c_2 d_3 = p_z \quad \ldots[7]$$

Solving, $$d_3 = s_2(c_1 p_x + s_1 p_y) = c_2 p_z \quad \ldots[8]$$

Now,

$$ {}^0_6T = \begin{bmatrix} n_x & s_x & a_x & p_x \\ n_y & s_y & a_y & p_y \\ n_z & s_z & a_z & p_z \\ 0 & 0 & 0 & 1 \end{bmatrix} = {}^0_3T\ {}^1_2T\ {}^2_3T\ {}^3_4T\ {}^4_5T\ {}^5_6T$$

$$\Rightarrow ({}^3_4T)^{-1}\ ({}^2_3T)^{-1}\ ({}^1_2T)^{-1}\ ({}^0_1T)^{-1}\ ({}^0_6T) = ({}^4_5T)\ ({}^5_6T)$$

$$({}^1_2T)^{-1} = \begin{bmatrix} c_2 & 0 & s_1 & 0 \\ -s_2 & 0 & c_2 & 0 \\ 0 & 1 & 0 & 0 \\ 0 & L_2 & 0 & 1 \end{bmatrix}$$

$$({}^2_3T)^{-1} = \begin{bmatrix} 1 & 0 & 0 & 0 \\ 0 & 0 & 1 & 0 \\ 0 & -1 & 0 & 0 \\ 0 & -d_3 & 0 & 1 \end{bmatrix}$$

$$({}^3_4T)^{-1} = \begin{bmatrix} c_4 & s_4 & 0 & 0 \\ -s_4 & c_4 & 0 & 0 \\ 0 & 0 & 1 & 0 \\ 0 & 0 & 0 & 1 \end{bmatrix}$$

Therefore, LHS $= ({}^3_4T)^{-1}\, ({}^2_3T)^{-1}\, ({}^1_2T)^{-1}\, ({}^0_1T)^{-1}\, ({}^0_6T) =$

$$\begin{bmatrix} x & x & c_4[c_2(c_1a_x + s_1a_y) - s_2a_z] + s_4(c_1a_y - s_1a_x) & x \\ x & x & -s_4[c_2(c_1a_x + s_1a_y) - s_2a_z] + c_4(c_1a_y - s_1a_x) & x \\ x & x & s_2(c_1a_x + s_1a_y) + c_2a_z & x \\ 0 & 0 & d_3[s_2(c_1a_x + s_1a_y) + c_2a_z] + L_2(c_1a_y - s_1a_x) & x \end{bmatrix}$$

$$\text{RHS} = {}^4_5T\ {}^5_6T = \begin{bmatrix} x & x & s_5 & x \\ x & x & 0 & x \\ x & x & c_5 & x \\ 0 & 0 & 0 & 1 \end{bmatrix}$$

Equating corresponding terms,

$$-s_4\,[c_2\,(c_1\,a_x + s_1\,a_y) - s_2\,a_z] + c_4\,(c_1\,a_y - s_1\,ax) = 0$$

$$\frac{s_4}{c_4} = \tan\,\theta_4 = \frac{c_1a_y - s_1a_x}{c_2(c_1a_x + s_1a_y) - s_2a_z}$$

$$\theta_4 = \tan^{-1}\left[\frac{c_1a_y - s_1a_x}{c_2(c_1a_x + s_1a_y) - s_2a_z}\right] \qquad \ldots[9]$$

$$s_5 = c_4\,[c_2\,(c_1\,a_x + s_1\,a_y) - s_2\,a_z] + s_4\,(c_1\,a_y - s_1\,a_x)$$

$$\text{and } c_5 = s_2\,(c_1\,a_x + s_1\,a_y) + c_2\,a_z$$

Therefore,

$$\tan\,\theta_5 = \frac{s_5}{c_6} = \frac{c_4[c_2(c_1a_x + s_1a_y) - s_2a_z + s_4(c_1a_y - s_1a_x)}{s_2(c_1a_x + s_1a_y) + c_1a_z}$$

$$\theta_5 = \tan^{-1}\left[\frac{c_4[c_2(c_1a_x + s_1a_y) - s_2a_z + s_4(c_1a_y - s_1a_x)}{s_2(c_1a_x + s_1a_y) + c_1a_z}\right] \qquad \ldots[10]$$

Again,

$$({}^4_5T)^{-1}\,({}^3_4T)^{-1}\,({}^2_3T)^{-1}\,({}^1_2T)^{-1}\,({}^0_1T)^{-1}\,({}^0_6T) = ({}^5_6T)$$

$$ {}^5_6T = \begin{bmatrix} c_6 & -s_6 & 0 & 0 \\ 0 & 0 & -1 & 0 \\ s_6 & c_6 & 0 & 0 \\ 0 & 0 & 0 & 1 \end{bmatrix}$$

$$\text{LHS} = ({}^4_5T)^{-1}\,({}^3_4T)^{-1}\,({}^2_3T)^{-1}\,({}^1_2T)^{-1}\,({}^0_1T)^{-1}\,({}^0_6T)$$

$$\begin{bmatrix} c_5\{c_4[c_2(c_1n_x+s_1n_y)-s_2n_z]+s_4(c_1n_x-s_1n_z)\}-[s_2(c_1n_x+s_1n_y)+c_2n_z]s_5 & x & x & x \\ -s_4[c_2(c_1n_x+s_1n_y)-s_2n_z]+c_4(c_1n_y-s_1n_x) & x & x & x \\ x & x & x & x \\ & x & x & x \end{bmatrix}$$

Equating with ${}^5_6T$'s corresponding terms

$$\theta_6 = \tan^{-1}\left[\frac{-s_4[c_2(c_1n_x+s_1n_y)-s_2n_z]+c_4(c_1n_y-s_1n_x))}{c_5\{c_4[c_2(c_1n_x+s_1n_y)-s_2n_z]+s_4(c_1n_y-s_1n_z)\}-[s_2(c_1n_x+s_1n_y)+c_2n_z]}\right] \quad \text{...[11]}$$

From (6) and (3)

i.e.,

$$[s_2d_3 = c_1p_x+s_1p_y]\,Xs_1$$
$$[s_2d_3 = c_1p_x+s_1p_y]\,Xs_1$$
$$[L_2 = -s_1p_x+c_1p_y]\,Xc_1$$

Adding, $L_1c_1+s_1s_2d_3 = p_y$ ...[12]

Again, multiplying (6) by c_1 and (7) by s_1 and subtracting,

$$p_x = c_1s_2d_3 - s_1L_2 \quad \text{...[13]}$$

And finally, from equation (7) we have $p_z = c_2d_3$

Inverse solution for velocity

Velocity components of the tool can be obtained by differentiating equations (7), (12) and (13) w.r.t. time:

$$P_z = c_2D_3 - s_2D_3\theta_2 \quad \text{...[14]}$$

$$P_y = -L_2s_1\theta_1 + s_1s_2D_3 + c_1s_2D_3\theta_1 + s_1c_2D_3\theta_2$$
$$= (-L_2s_1 + c_1s_2D_3)\theta_1 + s_1c_2D_3\theta_2 + s_1s_2D_3 \quad \text{...[15]}$$

$$P_x = c_1s_2D_3 + c_1c_2D_3\theta_2 - s_1s_2D\theta_1 - c_1L_2\theta_1$$
$$= -(s_1s_2D_3 + c_1L_2)\theta_1 + c_1c_2D_3\theta_2 + c_1s_2D_3 \quad \text{...[16]}$$

These equations can be solved to obtain the joint velocities (θ_1, θ_2, d_3) since we already know p_x, p_y, p_z. Here, P is the derivative of p, θ is **d(θ)/dt** and D_3 is **d(d$_3$)/dt**.

The remaining joint velocities (θ_4, θ_5, θ_6) can be obtained by differentiating once the equations for θ_4, θ_5 and θ_6 worked out in the previous section.

Inverse solution for acceleration:

By differentiating again i.e. (θ_1, θ_2, d_3, θ_4, θ_5, θ_6), we can examine the acceleration of the joints and by differentiating the equations of p_x, p_y and p_z, acceleration components of the tool or end effector can also be examined.

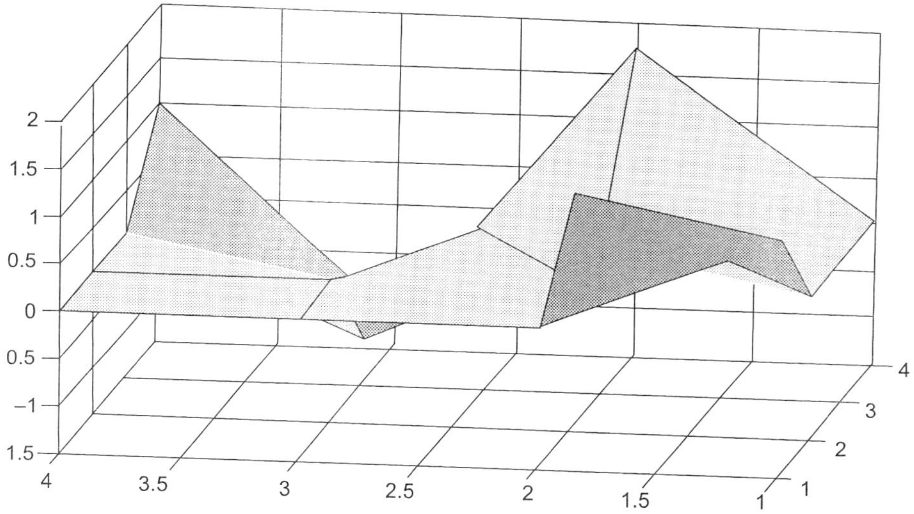

Figure 10.3: End Effector Simulation.

10.6 SIMULATION

Simulation was carried out for a fixed length of the second link equal to 1.75 m. In the algorithm written, the prismatic joint length was varied from zero m to 1.25 m in a loop with an increment of 0.01 m. The revolute links were varied from zero radian to 2 pi radians in for loops, with increments of pi/6 radians. The graph, which shows the end effector movement for the specified values, was obtained from the kinematic matrix of the forward kinematics.

Algorithm:

```
L2 = 1.75
for d3 = 0:0.01:1.25
end
for x1 = 0:pi/6:(2*pi)
end
for x2 = 0:pi/6:(2*pi)
end
for x4 = 0:pi/6:(2*pi)
end
```

```
for x5 = 0:pi/6:(2*pi)
end
for x6 = 0:pi/6:(2*pi)
end

K= [ cos(x6)*[cos(x5)*[cos(x1)*cos(x2)*cos(x4)–sin(x1)*sin(x4)]–
sin(x5)*cos(x1)*sin(x2)]–sin(x6)*[cos(x1)*cos(x2)*sin(x4)+sin(x1)*cos(x4)] –
sin(x6)*[cos(x5)*[cos(x1)*cos(x2)*cos(x4)–sin(x1)*sin(x4)]–sin(x5)*cos(x1)*sin(x2)]–
cos(x6)*[cos(x1)*cos(x2)*sin(x4)+sin(x1)*cos(x4)]sin(x5)*[cos(x1)*cos(x2)*cos(x4)–
sin(x1)*sin(x4)]+cos(x5)*cos(x1)*sin(x2) cos(x1)*sin(x2)*d3-sin(x1)*L2;
cos(x6)*[cos(x5)*[sin(x1)*cos(x2)*sin(x4)+cos(x1)*sin(x4)]–
sin(x5)*sin(x1)*sin(x2)]+sin(x6)*[cos(x1)*cos(x4)–sin(x1)*cos(x2)*sin(x4)] –
sin(x6)*[cos(x5)*[sin(x1)*cos(x2)*sin(x4)+cos(x1)*sin(x4)]–
sin(x5)*sin(x1)*sin(x2)]+cos(x6)*[cos(x1)*cos(x4)–sin(x1)*cos(x2)*sin(x4)]
sin(x5)*[sin(x1)*cos(x2)*sin(x4)+cos(x1)*sin(x4)]+cos(x5)*sin(x1)*sin(x2)
sin(x1)*sin(x2)*d3+cos(x1)*L2; cos(x6)*[cos(x2)*sin(x5)–
sin(x2)*cos(x4)*cos(x5)]+sin(x6)*sin(x2)*sin(x4) sin(x6)*[cos(x2)*sin(x5)–
sin(x2)*cos(x4)*sin(x5)]+cos(x6)*sin(x2)*sin(x4) –
[sin(x2)*cos(x4)*sin(x5)+cos(x2)*cos(x5)] –cos(x2)*d3; 0 0 0 1]
```

10.7 SUMMARY

A mathematical model of the Stanford Scheinmann an arm was developed and presented with simulation. Forward and Inverse Kinematics equations were derived using D-H notation. Simulation was performed using MATLAB. Using Forward Kinematics, Inverse Kinematics and its solutions for velocities and accelerations, robot parameters can be investigated which will facilitate the process of designing, construction and inspection of the robots in the real world. The simulation results are displayed in graphical format and motion of the end effector can be observed.

10.8 EXERCISE

Develop a mathematical model and perform simulation for a 4 degree of freedom robot using the approach given in this chapter. Select appropriate co-ordinates.

Appendices

APPENDIX-I

The governing equations for simulation (Chapter 4 – The NPD Model):

Acceptance rate = Work in Progress/quality level {Units: (person*Month)/Month}

Available time = IF THEN ELSE (scheduled completion time-Time > 0, scheduled completion time-Time, 1) {Units: Month}

Average salary = 500 {Units: dollars/ (Month*person) [0, 10000]}

Change in scope = 0 {Units: Dmnl [0, 1, 0.2]}

Change request time = 20 {Units: Month}

Compensation = INTEG (Salary, 0) {Units: Dollars}

Competence training = 1 {Units: Month [0, 1]}

Competent Project Staffs Pool = INTEG (training rate – quitting rate – loss in competence + gain in competence, 5) {Units: person}

Employee assigned for rework = IF THEN ELSE (Rework/rework time < Competent Project Staffs Pool, Rework/rework time, Competent Project Staffs Pool) {Units: person}

FINAL TIME = 36 {Units: Month}

Gain in competence = Incompetent Project Staffs Pool*competence training {Units: Dmnl/person}

Hiring factor = 1 {Units: Dmnl/Month [0, 1]}

Hiring rate = IF THEN ELSE (project end = 1, 0, quitting rate + IF THEN ELSE (Work to be done/Available time > (Competent Project Staffs Pool-employee assigned for rework), ((Work to be done/available time)-(Competent Project Staffs Pool-employee assigned for rework)*hiring factor, 0)) {Units: person/Month}

Incompetent Project Staffs Pool = INTEG (+ loss in competence – gain in competence, 0) {Units: person}

Increase in work normal = 0.2 {Units: Dmnl/Month [0, 1]}

INITIAL TIME = 0 {Units: Month – The initial time for the simulation.}

Initial work = 100 {Units: person*Month [0, 1]}

Lookup table factor = 1 {Units: Dmnl [0,100]}

Loss in competence = IF THEN ELSE (project end = 1, 0, IF THEN ELSE(change request time = Time, Competent Project Staffs Pool*change in scope, 0)) {Units: Dmnl/person}

Obsolescence normal = 1*change in scope {Units: Dmnl [0, 1]}

Obsolescence rate = IF THEN ELSE (Time = change request time, Work Finished*obsolescence normal, 0) {Units: person*Month/Month}

Potential Project Staffs Pool = INTEG (Hiring rate-training rate, 5) {Units: person}

Productivity = Productivity lookup (Competent Project Staffs Pool) {Units: Dmnl/person}

Productivity lookup ([(0,0) – (100, 1)], (0, 0), (0.01, 1), (1, 1), (10, 0.5455), (20, 0.531), (30, 0.5065), (40, 0.472), (50, 0.4275), (60, 0.373), (70, 0.3085), (80, 0.234), (90, 0.1495), (100, 0.055)) {Units: Dmnl/person}

Lookup is given according to work carried out by Fried (1995).

Project end = IF THEN ELSE (Work to be done < = 0: AND: Work Finished > = initial work, 1, 0) {Units: Dmnl}

Quality level = 1 {Units: Dmnl/Month [0.5, 1]}

Quitting rate = Competent Project Staffs Pool*schedule pressure*0.01 {Units: Dmnl/Month}

Rejection normal = 0.05*schedule pressure {Units: Dmnl/Month}

Rejection rate = Rejection normal*Work in Progress {Units: Dmnl}

Revenue = INTEG (Revenue generated-Compensation, 2e + 006) {Units: dollars}

Initial capital invested is 2 million US dollars.

Revenue generated = Work Finished*5000 {Units: dollars}

It has been assumed that for every unit of work finished, $5000 revenue is being generated.

Rework = INTEG(Obsolescence rate + rejection rate – rework rate, 0) {Units: person*Month}

Rework normal = 1 {Units: person*Month/Month [0, 1]}

Rework rate = IF THEN ELSE (Rework < = 0, 0, employee assigned for rework*rework normal) {Units: (person*Month)/Month}

Rework time = 1 {Units: Month}

Salary = IF THEN ELSE (project end = 1, 0, Competent Project Staffs Pool*average salary + (0.75*Potential Project Staffs Pool*average salary)) {Units: dollars/Month}

SAVEPER = TIME STEP {Units: Month [0, 36]}

The frequency with which output is stored.

Schedule pressure = Schedule pressure table lookup (available time/scheduled completion time) {Units: Dmnl}

Schedule pressure table lookup ([(0, 0.8) – (1, 2)], (0, 2), (0.1, 1.8), (0.15, 1.5), (0.25, 1.25), (0.5, 1.1), (1, 1)) {Units: Dmnl}

Scheduled completion time = 32 {Units: Month}

TIME STEP = 0.25 {Units: Month [0, 36]}

The time step for the simulation.

Training look up([(0, 0) – 100, 10)], (0, 0), (1, 1), (6, 2), (11, 3), (21, 4), (50, 5), (100, 5)) {Units: Month}

Training rate = IF THEN ELSE (project end = 1, 0, (Potential Project Staffs Pool/available time)*training time) {Units: person/Month}

Training time = Training look up (lookup table factor) {Units: Month}

Work Finished = INTEG (Acceptance rate – obsolescence rate, 0) {Units: person*Month}

Workflow normal = 1 {Units: persons [0, 1]}

Workflow rate = (Work to be done*workflow normal)/ (((Competent Project Staffs Pool-employee assigned for rework)*productivity) + (Work in Progress))

{Units: person*Month/Month}

Work in Progress = INTEG (– rejection rate + workflow rate + rework rate – acceptance rate, 0) {Units: person*Month}

Work rate = IF THEN ELSE (Time = change request time, change in scope* initial work*increase in work normal, 0) {Units: person*Month/Month}

Work to be Done = INTEG (+ work rate – workflow rate, Initial work)

{Units: person*Month}

APPENDIX-II

Equations: Insurance Sector Dynamics

1. core service = underpinning knowledge, 100
2. underpinning knowledge = core service*rate of service knowledge gain
3. rate of service knowledge gain = 0.1
4. human element of service = manpower*productivity
5. manpower = 100
6. productivity = 0.6
7. non-human element = efficiency of personnel*ICT tools
8. ICT tools = 10
9. efficiency of personnel = 0.6
10. tangibles of service = knowledge
11. knowledge = rate of knowledge gain*tangibles of service
12. rate of knowledge gain = 0.1
13. social responsibility = satisfaction index*societal needs
14. societal needs = 10
15. satisfaction index = 0.6
16. service quality perception = core service + human element of service + non-human element + social responsibility + tangibles of service
17. external communication = customer contacts, 100
18. customer contacts = contact rate*external communication
19. contact index = 1
20. word of mouth = contact index*number of insurance holders
21. number of insurance holders = 100
22. contact index = 1
23. past experience = influence factor*customer perception of expectation
24. influence factor = 0.6
25. customer perception of expectation = 3000
26. personal needs = customer perception of expectation*personal need factor
27. personal need factor = 0.6
28. external communication = customer contacts,100
29. customer contacts = contact rate* external communication
30. contact rate = 0.1
31. service quality expectation = past experience + personal needs + external communication + word of mouth
32. service quality gap = service quality perception – service quality expectation

APPENDIX-III

Equations: Manufacturing Supply Chain

(01) Actual Quality = 0.9
Units: Dmnl

(02) Adaptation to Production Rate = (Desired Distribution Inventory – Distribution Inventory)/ Adjustment Time
Units: Units/Month

(03) Adaptation to Production Start Rate = (Desired Work in Progress – Work In Progress)/ Production Cycle
Units: Units/Month

(04) Adjustment Time = 1
Units: Month

(05) Available Inventory = 1700
Units: Units/Month

(06) Capacity of Each Machine = 50
Units: Units/Month

(07) Change in expected output = – (Expected Inventory Output – Inventory Output Rate)/Delay
Units: Units/(Month*Month)

(08) Delay = Maintenance + Shipments from Suppliers
Units: Month

(09) Demand = Forecasts*Quality/Price
Units: Units/Month

(10) Desired Distribution Inventory = Expected Inventory Output*Desired Inventory Coverage
Units: Units

(11) Desired Inventory Coverage = 1
Units: Month

(12) Desired Production Rate = Adaptation to Production Rate + Expected Inventory Output + (MIN("No. of Machines", "No. of Workers")*Productivity*Capacity of Each Machine)
Units: Units/Month

(13) Desired Production Start Rate = Desired Production Rate + Adaptation to Production Start Rate
Units: Units/Month

(14) Desired Work In Progress = Desired Production Rate*Production Cycle
Units: Unit

(15) Distribution Inventory = INTEG (Production Rate – Distribution Rate – Waste Rate, 1700)
Units: Units

(16) Distribution Rate = MAX(Demand, Distribution Inventory)
Units: Units/Month

(17) Expected Inventory Output = INTEG (Change in expected output, 1700)
Units: Units/Month

(18) FINAL TIME = 60
Units: Month
The final time for the simulation.

(19) Forecasts = 3.86153e + 006
Units: Rs/Month

(20) Idle Time = 0.1
Units: Month

(21) INITIAL TIME = 0
Units: Month
The initial time for the simulation.

(22) Inventory Output Rate = Distribution Rate + Waste Rate
Units: Units/Month

(23) Maintenance = 2
Units: Month

(24) No. of Machines = 10
Units: Dmnl

(25) No. of Workers = 10
Units: Dmnl

(26) Normal Storage Cost = 0.1
Units: Rs/(unit*time)

(27) Price = 205
Units: Rs/Unit

(28) Production Cycle = 3
Units: Month

(29) Production Rate = Work in Progress/Production Cycle
Units: Units/Month

(30) Production Start Rate = Desired Production Start Rate –Available Inventory
Units: Units/Month

(31) Productivity = Productivity of Machines *Productivity of workers
Units: Dmnl

(32) Productivity of Machines = 0.9
Units: Dmnl

(33) Productivity of workers = 0.8
Units: Dmnl

(34) Quality = Actual Quality*Quality Perceptive Index
Units: Dmnl

(35) Quality Perceptive Index = 0.8
Units: Dmnl

(36) Rework Time = 0.2
Units: Month

(37) SAVEPER = TIME STEP
Units: Month
The frequency with which output is stored.

(38) Shipments from Suppliers = 2
Units: Month

(39) Storage Cost = 0.1
Units: Rs/(unit*time)

(40) TIME STEP = 1
Units: Month
The time step for the simulation.

(41) Waste Rate = Distribution Inventory/Waste Time*(Normal Storage Cost/Storage Cost)
Units: Units/Month

(42) Waste Time = Maintenance + Rework Time + Idle Time
Units: Month

(43) Work in Progress = INTEG (Production Start Rate –Production Rate, 6300)
Units: Unit.

APPENDIX-IV

Equations: Innovation Dynamics in Manufacturing

(01) Brand image = INTEG (Market performance, 20)
Units: units

(02) Capacity of innovation = INTEG (Innovation success, 0)
Units: units

(03) Competence = Knowledge through combination + Knowledge through externalization + Knowledge through internalization + Knowledge through socialization
Units: units

(04) Customer volume = INTEG (New customers, New customers)
Units: units

(05) FINAL TIME = 120
Units: Month
The final time for the simulation.

(06) INITIAL TIME = 0
Units: Month
The initial time for the simulation.

(07) Innovation = (Competence + Knolwledge gained through feedback)*Productivity factor
Units: units

(08) Innovation success = (Rate of innovation success*Rate of knowledge transfer)+Innovative Ability
Units: units

(09) Innovative Ability = INTEG (Innovation, 10)
Units: units

(10) Knolwledge gained through feedback = Useful knowledge factor*Customer volume*0.01/10^10
Units: units

(11) Knowledge = Total Knowledge*Redundancy factor
Units: units

(12) Knowledge through combination = 1000*(1 + Rate of Knowledge gain through combination)
Units: units

(13) Knowledge through externalization = (1000*Rate of Knowledge gain through externalization)
Units: units

(14) Knowledge through internalization = 1000*(1 + Rate of Knowledge gained through internalization)
Units: units

(15) Knowledge through socialization = 1000*(1 + Rate of Knowledge gained through socialization) + Knolwledge gained through feedback
Units: units

(16) Market performance = Capacity of innovation*Personal needs
Units: units

(17) New customers = Potential customers*Brand image
Units: units

(18) Organizational knowledge = INTEG (Knowledge, Knowledge)
Units: units

(19) Personal need factor = 0.25
Units: units [0, 1, 0.05]

(20) Personal needs = Personal need factor
Units: units

(21) Potential customers = 500
Units: units [0, 40000, 500]

(22) Productivity factor = 0.2
Units: units [0, 1, 0.05]

(23) Rate of innovation success = 0.15
Units: units [0, 1, 0.05]

(24) Rate of knowledge gained through combination = 0.1
Units: units [0, 1, 0.05]

(25) Rate of knowledge gained through externalization = 0.1
Units: units [0, 1, 0.05]

(26) Rate of knowledge gained through Internalization = 0.1
Units: units [0, 1, 0.05]

(27) Rate of knowledge gained through socialization = 0.1
Units: units [0, 1, 0.05]

(28) Rate of knowledge transfer = 0.4*Organizational knowledge
Units: units

(29) Redundancy factor = 10
Units: units [0, 1, 0.1]

(30) SAVEPER = TIME STEP
Units: Month [0, 10]
The frequency with which output is stored.

(31) TIME STEP = 1
Units: Month [0, 10]
The time step for the simulation.

(32) Total Knowledge = Knowledge through combination + Knowledge through externalization + Knowledge through internalization + Knowledge through socialization
Units: units

(33) Useful knowledge factor = 0.4
Units: units [0, 1, 0.05]

References

- AIAA. (1998), "Guide for the Verification and Validation of Computational Fluid Dynamics Simulations", American Institute of Aeronautics and Astronautics, www.aiaa.org/store, (Last accessed, 17 October 2005).
- Almor, T. & Hashai, N., (2004), "The Competitive Advantage and Strategic Configuration of Knowledge-intensive, Small and Medium-sized Multinationals: a Modified Resource-based View", Journal of International Management, Vol. 10, pp. 479-500.
- Barlas, Y. (1989), "Tests of model behaviour that can detect structural flaws: demonstrations with simulation experiments", Computer Based Management of Complex Systems, P.M. Milling and K. Zahn. Germany, Springer-Verlag, pp. 246-254.
- Barlas, Y. (1996), "Formal aspects of model validity and validation in system dynamics", System Dynamics Review, 12(3), pp. 183-210.
- Barney, J.B., (1991), "Firm Resources and Sustained Competitive Advantage", Journal of Management, Vol. 17, pp. 99-120.
- Belliveau, P., Griffin, A. and Somermeyer, S. (2002). *PDMA Tool Book 1 for New Product Development*, John Wiley, New York, 2002.
- Bitner, M.J. (1992). "Servicescapes: the impact of physical surroundings on customers and employees," Journal of Marketing, Vol. 56, pp. 57-71.
- Carson, J.S. (1989), "Verification and Validation: A Consultant's Perspective", E.A. MacNair, K.J. Musselman, and P. Heidelberger (Eds.), Proc. 1989 Winter Simulation Conf, pp. 552-557.
- Cohen, W. & Levinthal, D., (1990), "Absorptive Capacity: A new perspective on learning and Innovation", Administrative Science Quarterly, Vol. 35, pp. 128-152.
- Coyle, R.G. (1977), "Management System Dynamics", John Wiley and Sons, Chichester.
- Craig, John J., (1986). "Introduction to Robotics Mechanics and Control", MA: Addison-Wesley.
- DMSO (Defence Modelling and Simulation Office). (1996), "The Principles of Verification, Validation, and Accreditation", Verification, Validation, and Accreditation Recommended Practices Guide, www.dmso.mil/public, U.S. Department of Defense, Office of the Director of Defence Research and Engineering, November (Last accessed, 17 October 2005).
- Forrester, J.W. (1958), "Industrial Dynamics: A Major Breakthrough for Decision Makers", Harvard Business Review, vol. 36, no. 4, pp. 37-66.
- Forrester, J.W. and Senge, P. (1980), "Tests for building confidence in System Dynamics Models", TIMS Studies in the Management Sciences, 14, pp. 209-228.
- Fried, Louis, (1995) "Managing Information Technology in Turbulent Times." John Wiley, New York.

- Giannanasi, F., Lovett, P. and Godwin, A.N. (2001), "Enhancing confidence in discrete event simulations", Computers in Industry, 44, pp. 141-157.
- Goldberg, J., Dietrich, R., Chen, J., Valenzuela, T., Criss. E. (1990), "A Simulation Model for Evaluating a Set of Emergency Vehicle Base Locations: Development, Validation, and Usage", Socio-Economic Planning Science, 24, pp. 125-141.
- Ijeoma, S.I., Andersson, J. & Wall, A. (2001), "Correctness criteria for models' validation-A philosophical perspective", Department of Computer Science & Computer Engineering (IDT), www.mrtc.mdh.se/publications/0731.pdf, Malardalen University, Sweden (Last accessed, 17 October 2005).
- Kotler, P., (1973). "Atmospherics as a marketing tool", Journal of Retailing, vol. 49, pp. 48-64.
- Law, A.M. and McComas, M.G. (2001), "How to Build Valid and Credible Simulation Models", Proceedings of the 2001 Winter Simulation Conference, (Eds.), B.A. Peters, J.S. Smith, D.J. Medeiros, and M.W. Rohrer, pp. 22-29.
- Law, A., M. and Kelton, W., D. (1991). Simulation Modelling & Analysis, 2nd ed., McGraw-Hill, pp. 306.
- Lee, H.L., P. Padmanabhan, and S. Whang, (1997), "Information Distortion in a Supply Chain: The Bullwhip Effect," Management Science vol. 43, pp. 546-58.
- Love, G. and Back, G. (2000), "Model Verification and Validation for Rapidly Developed Simulation Models: Balancing Cost and Theory", Proceedings of the 18th International Conference of the System Dynamics Society, Bergen, Norway, August 6-10.
- McKeown, Max (2008). The Truth About Innovation. London, UK: Prentice, ISBN 0273719122.
- Mohapatra, P.K.J. (1987), "Validation of System Dynamics Models", Orientation Course, Lecture Notes, Second National Conference on System Dynamics, Varanasi, January, 15-17.
- Nonaka, I. & Takeuchi, H. 1995, 'The Knowledge Creating Company', Oxford University Press, New York.
- Panov, S.A. and V.I. Shiryaev (2003), "Manufacturing supply chain adaptive management under changing demand conditions", System Dynamics Conference Proceedings, MIT Boston.
- Parasuraman, A, Zeithaml, V.A and Berry, L.L. (1988). "SERVQUAL: a multiple-item scale for measuring consumer perceptions of service quality", Journal of Retailing, Spring, pp. 12-40
- PMBOK, (2000). "A guide to the project management book of knowledge" (PMBOK® guide), 2000 ed.
- Richardson, G.P. and Pugh III, A.L. (1981), "Introduction to System Dynamics Modelling with DYNAMO", The MIT Press, Cambridge, Massachusetts.
- Saysel, A.K. and Barlas, Y. (2004), "Model Simplification and Validation: Illustration with Indirect Structure Validity Tests", Working Papers in System Dynamics, March, ISSN 1503-4860.

- Schlesinger, S. (1979), "Terminology for Model Credibility", Simulation, 32(3), pp. 103-104.
- Solberg, J. (1992), "The power of simple models in manufacturing", Manufacturing systems– Foundations of World-Class Practice. J. Hein and W. Compton. USA National Academy of Engineering Press, Washington, pp. 215- 223.
- Spong, M.W., Hutchinson, S., Vidyasagar, M., (2004). "Robot Dynamics and Control", John Wiley & Sons, New York.
- Sterman, J. (1984), "Appropriate Summary Statistics for Evaluating the Historic Fit of System Dynamics Models", Dynamica, 10, winter, pp. 51-66.
- Sterman, J.D. (2000). "Business Dynamics: Systems Thinking and Modelling for a Complex World", Irwin McGraw-Hill, p. 845.
- Sterman, J.D. (2000). Business Dynamics: Systems Thinking and Modelling for a Complex World, Irwin McGraw-Hill, pp 845.
- Sureshchandar G.S., Chandrashekaran Rajendran, Anantharaman R.N. (2002) "Determinants of customer – perceived service quality: a confirmatory factor analysis approach", Journal of services marketing , vol.16, no1, pp. 9-34.
- Sushil, (1993), "System Dynamics: A Practical approach for Managerial Problems", Wiley Eastern Publication, New Delhi, ISBN: 81-224-0498-7, p. 37.
- Towill D., R. (1982). "Dynamic analysis of an inventory and order based production control system". International Journal of Production Research, 20, pp. 671-687.
- Zack, M.H., (1999a), 'Developing a Knowledge Strategy', California Management Review, vol. 41, no. 3, pp. 125-145.
- Zack, M.H., (1999b), 2000, Competing on Knowledge, Handbook of Business Strategy (New York: Faulkner & Gray), pp. 81-88.
- Zeithaml, V.A., Parasurarnan, A., and Berry, L.L. (1985). "Problems and strategies in services marketing," Journal of Marketing, vol. 49, pp. 33-46.

Index